Worthington Hooker

**Science for the School and Family**

Worthington Hooker

**Science for the School and Family**

ISBN/EAN: 9783337034634

Printed in Europe, USA, Canada, Australia, Japan

Cover: Foto ©Paul-Georg Meister /pixelio.de

More available books at **www.hansebooks.com**

# SCIENCE

## FOR THE

# SCHOOL AND FAMILY.

### PART III.
### MINERALOGY AND GEOLOGY.

BY

WORTHINGTON HOOKER, M.D.,

PROFESSOR OF THE THEORY AND PRACTICE OF MEDICINE IN YALE COLLEGE,
AUTHOR OF "HUMAN PHYSIOLOGY," "CHILD'S BOOK OF NATURE,"
"NATURAL HISTORY," &c.

*Illustrated by nearly Two Hundred Engravings.*

NEW YORK:
HARPER & BROTHERS, PUBLISHERS,
FRANKLIN SQUARE.
1865.

# PREFACE.

This book is intended to meet the wants of *beginners* in the study of Geology, and especially young beginners. There are many elementary books, so called, on this science, but they all, so far as I have seen them, contain much matter which is suited only to those who have already become acquainted with the subject. They cover too much ground for a beginner. And, besides this, they are not sufficiently simple in their explanations and illustrations. These defects I have endeavored to avoid in the construction of this book. My object has been to produce a text-book fitted to prepare those who are wholly unacquainted with the subject for the farther study of it in the books of professed geologists. I would gladly have left this task to some of them, but, as no one has met the existing want, I have undertaken to do it, though with some hesitation, from my lack of familiarity with the full minutiæ of the subject. But perhaps the fact that I have been obliged to be, to some extent, a learner, in order to accomplish my task, has the better fitted me for it, as I have thus become sensible of the wants of the learners for whom I write.

As in my Chemistry (Part II.), so in this book, I have made it a point to convey to the pupil simply that knowledge of the study which *every well-informed person ought to possess*, leaving out those minutiæ which are of value only to one who intends to be a thorough geologist. I have also brought out very prominently those common geological phenomena which are within the scope of ordinary observation, so that the pupil may be prompted, by what he learns in this book, to prosecute his observations and investigations in his daily walks, and in his travels to different localities, far or near.

I do not go largely into Mineralogy, but present enough of it simply to prepare the pupil for the study of Geology. Observe in this connection the natural succession of subjects in the three parts of this portion of my series. The Natural Philosophy properly precedes the Chemistry, and this latter the Mineralogy and Geology, each study preparing the pupil to understand what comes after. It should be remarked here that these studies, together with that of Zoology, should precede the study of Physical Geography, for acquaintance with them is absolutely essential to any thing like a full knowledge of that science; and yet it is very common to see classes wearily plodding through some book on Physical Geography, who, from lack of this preparation, find most of it a *terra incognita*, and know but little of what they have passed through when they arrive at their journey's end. Let me not be understood to say that this science should not be studied at all till after the pupil has gone through with the studies that I have mentioned. It is only the full consideration of it that should be delayed till that period. It should be introduced, more or less, throughout the whole previous course, mingled with the teaching of Geography as it is now ordinarily pursued. But this can not be done properly and effectually unless the study of the natural sciences be made a part of education from the outset, as is contemplated in my series of books, beginning with the "Child's Book of Common Things," and ending with the present work. That this view of the matter is correct is readily seen if you observe how tributary are the natural sciences to the interest and value of the study of Geography, for without their contributions to it this branch is nothing more than mere dry topography and statistics.

The materials for this book I have drawn from various sources. Foremost stand the works of Professor Dana, and especially that grand American book, his Manual of Geology. Then there are the books of Lyell, Hitchcock, Hugh Miller, Phillips, Gray and Adams, Richardson, etc. I mention with special pleasure some smaller works from the English press. "The Past and Present Life of the Globe," by Page, is a most able and interesting commentary on the life-record of the earth's formation. "Geology in the Garden," by Rev. Henry Eley, shows in a very ingenious manner what may be learned about English Geology by following out the suggestions derived from the examination of the soil of a garden. One of the best books for a beginner that I have met with is the "School Manual of Geology," by Mr. Jukes, local director of the geological survey of Ireland. It would be well for the teacher to have some of the books which I have mentioned for the purpose of reference, so that he may, if he wishes, bring up additional matter in his teaching, or answer any inquiries of his pupils on points that may not be fully treated in this book. This would be in consonance with the principles I laid down in the preface to my Natural History in regard to the relative use of text-books and books for reference.

I may say of this book as I said of Part II., that in the present state of education it is adapted to high schools and academies; and yet it would be within the grasp of the older pupils of our common schools if they had studied in their proper time all the other books of the series, for this is no greater height than is reached in some other branches, where all the steps of the gradation are taken from the beginning, as, for example, in Mathematics.

That such studies as are provided for in the series, of which this is the concluding book, should be pursued, to some extent, before arriving at the high school or the academy, is very clear, if you consider the fact that the majority of pupils end their education when they pass from the higher classes of the common school. Most of these engage in various arts and trades, and so have to do constantly in the business of their life with the principles that are learned in these studies. Why, then, should not these principles be taught to them, it is pertinent to ask, if it be the great object of education to prepare one to act well his part in life? And in the case of those who are not to pursue any art in which the principles of natural science are brought into play, a knowledge of the phenomena that abound around us in air, water, and earth is not only valuable as information, and as a source of enjoyment, but also as a means of mental culture.

The questions in this and the previous parts I have put at the end of the book, believing that when they are nearer to the text both pupil and teacher are apt to depend on them too much. For this reason also I have the numbers refer to the pages, while in the Index I have the more definite reference to the paragraphs. In the Glossary there are no definitions, but references to paragraphs where the explanation of the terms can be found. W. HOOKER.

*January*, 1865.

# CONTENTS.

| CHAPTER | | PAGE |
|---|---|---|
| I. | MINERAL SUBSTANCES | 9 |
| II. | CONSTRUCTION OF MINERALS | 14 |
| III. | CARBON AND ITS COMPOUNDS | 23 |
| IV. | SULPHUR AND ITS COMPOUNDS | 33 |
| V. | METALS AND THEIR ORES | 39 |
| VI. | OXY-SALTS AND HALOID SALTS | 49 |
| VII. | EARTHY MINERALS | 56 |
| VIII. | ROCKS | 67 |
| IX. | THE EARTH AS IT IS | 79 |
| X. | PRESENT CHANGES IN THE EARTH | 90 |
| XI. | CONSTRUCTION OF THE EARTH | 123 |
| XII. | RECORD OF LIFE IN THE ROCKS | 160 |
| XIII. | AZOIC AGE | 181 |
| XIV. | AGE OF MOLLUSKS | 187 |
| XV. | AGE OF FISHES | 201 |
| XVI. | AGE OF COAL | 213 |
| XVII. | AGE OF REPTILES | 236 |
| XVIII. | AGE OF MAMMALS | 261 |
| XIX. | AGE OF MAN | 305 |
| XX. | CONCLUDING OBSERVATIONS | 314 |

# MINERALOGY AND GEOLOGY.

## CHAPTER I.

### MINERAL SUBSTANCES.

1. **Different Forms of Minerals.**—As solid substances alone are kept in mineral cabinets, minerals are thought of by most people as being only in the solid form. But there are mineral liquids and gases. Mercury is a liquid mineral. Water is a liquid composed of two mineral gases, oxygen and hydrogen. The atmosphere is a mixture of three mineral gases—oxygen, nitrogen, and carbonic acid, and holds always in solution some of a mineral liquid—water. All minerals are, in one sense, solid, for the atoms of which liquids and gases are composed are solid. All matter is undoubtedly, as Newton supposed, formed of "solid, massy, hard, impenetrable particles" (Part I., § 14).

2. **Relation of Heat to the Forms of Matter.**—The form which many mineral substances assume is dependent upon the degree of temperature to which they are exposed. Thus water appears in the three forms—solid, liquid, and gaseous, according to the degree of heat. We ordinarily speak of it as a liquid substance, because under all ordinary circumstances it has this form; but there are localities, as the tops of some mountains and the extreme polar regions, where its ordinary condition is that of a solid. For the same reason, we speak of metals, with the exception of mercury, as solids; but there was a time, in ages long gone by, as you will see in the geological part of this book, when these metals

were in a liquid state, and, with such a degree of heat as prevailed then, the water must have been in the vaporous form. Some of the gases have been reduced to a liquid and even a solid state by the combined influence of cold and pressure, and some have been made liquid by cold alone.

**3. Substances which are Not Mineral.**—Those substances are not considered mineral which have been produced by living agencies. They are of three kinds: 1. Organized substances, or those which are found in the structures of animals and vegetables. These are called organized because they have organs in them, as sap-vessels, blood-vessels, etc. Their mode of growth or increase is different from that of minerals, for minerals increase by additions wholly upon the *outside*, while organized substances grow at every point, internally as well as externally, by the circulation in their organs. 2. Substances contained in living organs, being formed by them, as starch, sugar, fats, etc. 3. Substances which come from certain changes in those of the first and second classes, as, for example, alcohol produced from sugar. For a more full exposition of this subject I refer you to Chapter XIX. of Part II.

**4. All Matter Once Mineral.**—There was a time when there was no life on this earth of ours, and, therefore, all the matter was mineral. This is indicated in the Mosaic account of the creation, for vegetable and animal life were introduced after certain preparations had been made for them; and what is thus indicated has been demonstrated to be true by the examination of the rocks by geologists, as you will see hereafter. This lifeless period occupied a long series of ages. All this time no organic substances were formed, for life always presides over their formation, using, indeed, such agencies as heat, light, and chemical action, but directing and controlling them. (Part II., § 509.)

**5. Life Introduced by the Creator.**—When the earth

was brought to a proper state for the agencies of life, then, and not till then, was life introduced by the Creator upon the stage of action. Then the mineral matter in earth, water, and air began to be acted upon by the new agencies, and there were evolved living forms, vegetable and animal. And these forms were capable of producing substances which no mere chemical action of the mineral atoms was able to produce during all the long lifeless ages of the world, or during the long ages since. The elements of which starch, sugar, fat, etc., are composed, have been in existence ever since the matter of our world was first created, and they exist now abundantly in the air and water; and yet, though they have always been acting upon each other, never have they been able so to combine as to produce these substances, unless life come in to direct their action. Life is continually evolving new forms and substances from dead mineral matter, which in the change it endows with properties that it did not before possess.

6. **Decay.**—In what is termed the decay of animal and vegetable substances, we have the opposite of the processes alluded to in § 5. Here we have a return of living substances to the state of dead mineral matter. There is no actual decay—that is, no real loss of matter, but merely change in the relations of the atoms which compose the substances. Life has let go its control, and the atoms obey the common laws of chemistry. For example: an egg has life in it, and that life, with the aid of a proper degree of heat, evolves within that prison of chalk a complicated living form—a bird; but let that life in some way be destroyed, and chemical action at once begins its work, returning the matter to the mineral world by combinations of its atoms in new forms. One of these combinations, for example, is a gas, which is the cause of the peculiarly disagreeable odor of a decaying egg—sulphureted hydrogen, resulting from the union of sulphur with hydrogen gas.

**7. Mineral Matter in Vegetables and Animals.**—Organic substances are not wholly destitute of mineral matters, and some have large quantities of such matter in them. There is silica or flint in plants, especially in the grasses; but this is one of the most extensively diffused of minerals, making up most of the sand of the earth, entering largely into the composition of granite and other rocks, and being more or less mingled up with the common earth every where. Then there is carbonate of lime, the mineral which constitutes the limestones, the chalks, and the marbles, making up the shells of the shell-fish and the skeletons of the coral animals; and phosphate of lime is the chief constituent of bone. But, in these and all other similar cases, there is mere deposition of mineral matter in interstices in the living organs, and no real change of it into living matter.

**8. Rocks Formed from the Mineral Substances in Animals.**—Quite a large portion of the rocks in the earth have been formed from the mineral remains of animals. It is supposed by some geologists that this is true of all which are composed of carbonate of lime. It is at least extensively true of them, as we know by the remains of shells and corals found in them. Much, if not all, the marble in the world was once a conglomerate of such remains, and was crystallized into its beautiful granular condition chiefly by the agency of heat. So, also, some flinty rocks are formed from the silica of the shields or skeletons of animals and vegetables, for the most part exceedingly minute in size. It is thus that these living forms gather from the water in which they live material for their structure, which, when they die, is laid down to be consolidated into rock. This interesting subject, barely touched here, will be fully treated of hereafter.

**9. Minerals Simple and Compound.**—Some minerals are found in nature in their simple elementary state alone, while others are found only in combination with other elements, and others still are found in both conditions. I will give some examples of each. Some of the noble metals (so called), as gold and platinum, are never found combined with any other element. On the other hand, the metals potassium, sodium, calcium, magnesium, aluminum, and ammonium are never found uncombined. Silver, copper, and mercury are among the examples of metals that are found in their simple state, or, as it is commonly expressed, found *native*, and also combined with other elements, as sulphur, oxygen, etc., in the form of ores. While oxygen is, in its simple state, in the mixture which we call air, it also appears in combination with almost all the elements, and forms from one third to one half of the crust of the earth. Hydrogen, which is so abundant as one of the elements of water, is never found uncombined. The same may be said of chlorine, one of the elements of common salt. Sulphur and carbon are both found native as well as combined, but phosphorus is always in combination.

10. **Minerals that Contain Many Elements.**—There is very great difference in degree in the compound character of the minerals found in mineralogical cabinets. While there are many of those compounds which were brought to your notice in Part II.—oxyds, sulphurets, salts, etc.—there are also many which are much more compound than these. The only approach made to such complex compounds by the processes of chemistry is in the double salts, such as alum, and the tartrate of antimony and potash. As examples, I mention mica, called isinglass by the common people, which is composed of silica, alumina, oxyd of iron, fluoric acid, and water; and that splendid mineral, so much used for making costly vases and other ornamental articles, lapis-lazuli, which is composed of silica, alumina, soda, lime, iron, sulphuric acid, sulphur, chlorine, and water.

11. **Mixtures of Compounds in the Rocks.**—In many of the rocks there are *mixtures* of the mineral compounds.

Granite is in contrast with marble in this respect. While the latter is composed of one compound mineral, the carbonate of lime, the former has mingled together in a confused manner three compounds—mica, feldspar, and quartz. The eye can generally distinguish readily the three kinds of crystals imperfectly formed. A full consideration of the composition of the rocks will claim our attention in another chapter.

12. **Mineralogy, Geology, and Chemistry.**—While Mineralogy has an intimate relation with Chemistry, it differs from it. It gets from Chemistry the composition of the various minerals of which it treats, but it has nothing to do, as Chemistry has, with what can be made out of these substances, or what can be extracted from them. It treats of substances simply as they are found in nature, without regard to any action that might be induced by mingling them together. Mineralogy is preparatory to Geology, because it gives us a knowledge of the mineral substances which make up the rocks. The three sciences are thus, as you see, linked together, and the proper order for their study is that which I have adopted.

## CHAPTER II.

### CONSTRUCTION OF MINERALS.

13. **Crystallization.**—When a mineral substance takes on a solid form, the atoms or particles are disposed to an arrangement which is termed crystallization. It is a very definite and exact arrangement, with straight lines, perfect angles, and plane, smooth faces. Crystals of quartz, commonly called rock crystal, are familiar examples. Mica is another of quite a different kind, the crystals being foliated—that is, in leaves. When the process of forming the crystal is not interfered with by any circumstance, as, for example, agitation, all parts of it

CONSTRUCTION OF MINERALS. 15

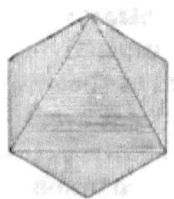

Fig. 1.

are perfect. The crystal of alum, for instance, is a perfect double pyramid, the shape being given in Fig. 1, which is an *octahedron* — that is, a body with eight equilateral angles, the lines all being exactly equal, and the spaces also. The name is from two Greek words, *octo*, eight, and *hedra*, base. In Fig. 2 you have the crystals of alum as

Fig. 2.

they appear when their formation is interfered with by some disturbing cause. In the glistening white marble the crystals are so crowded together as they form that none of them are perfect. The same is true of the crystals of quartz, mica, and feldspar in granite.

14. **Crystals of Different Sizes.**—As crystals increase by additions upon the outside, the crystals of any substance may vary greatly in size. Quartz crystals are sometimes very large. There is one at Milan which is $3\frac{1}{4}$ feet long, $5\frac{1}{2}$ in circumference, and weighs 870 pounds. Of course, the larger crystals of any substance will have precisely the same shape with the smaller.

**15. Crystallization and Vital Growth Contrasted.**—While crystallization adds to the outside alone, in living growth there is addition made to every part of the substance. The additions in the case of the crystal are invariably the same, and all parts of it are alike; but a living growth, whether vegetable or animal, commonly differs much in its different parts. A finger and a quartz crystal, for example, differ widely in this respect. Farther: while crystallization tends to straight lines and exact angles, the growth of animal and vegetable substances tends to curved lines, and its angles, when it makes any, are not sharply defined. The branches of trees are more or less rounded, and, while they have a general angular arrangement with the trunk of the tree, the angles are not definite.

**16. Exceptions.**—While what I have said of crystallization is generally true, there are some exceptions. The

Fig. 3.

faces of diamonds are commonly convex instead of plane, and the edges are therefore curved. In Fig. 3 you have the usual form of spathic iron (carbonate of iron) and pearl spar (magnesian carbonate of lime). There is sometimes seen in limestone, in clay-stones, and in some other rocks, a disposition to gather in spherical forms, and the process is at least akin to crystallization. The *arrangements* of crystals are very often in beautiful and varied imitation of branches, leaves, and flowers. This is very familiar to us in the frostings on our windows. This is seen in other minerals, as in alabaster in the Mammoth Cave of Kentucky, where leaves, vines, and flowers are imitated, some of the "rosettes" being, as stated by Professor Dana, a foot in diameter. In all such cases the curvings are in the arrangements of the crystals, each individual crystal probably being in its usual form.

**17. Mineral Matter in Living Substances.**—This subject has been already spoken of in § 7. I introduce it again here to notice the fact that the mineral matter is never deposited in living substances in any thing like a crystalline form. Though there is much silica (quartz) in the rushes, there are no quartz crystals; and the phosphate of lime in bones, though its particles are deposited after a certain definite plan (as seen in Fig. 84 in my "Human Physiology"), exhibits no resemblance to the crystals of this mineral.

18. **Crystallization not Confined to Minerals.**—Though organized substances, vegetable or animal, never take on a crystalline form, yet some of the products contained within them may do so. This is the case with the sugar of the sugar-cane and other plants. The "candying" of raisins is an example of the crystallization of sugar. The crystals which sugar is disposed to form are of the shape seen in Fig. 4, a six-sided prism, as you may observe in what is called rock-candy. In the form of loaf sugar the crystals are huddled together, and are therefore imperfect, just as in the case of marble. Many of the vegetable alkaloids, morphine, caffein, etc. (§ 590, Part II.), are obtained in the crystalline form. In all these cases the crystals are never formed so long as the substance remains under the influence of the living agency. Sugar, for example, never crystallizes in the living plant, but only after it is taken from it.

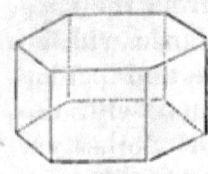

Fig. 4.

19. **Modes of Crystallization.**—There are various modes in which crystallization occurs. 1. One of the most common modes is by deposit from a solution. Thus alum, salt, sugar, etc., may form crystals from a solution, the water passing off by evaporation. When the substance can be dissolved in larger amount by hot water than by cold, as is the case with alum, but not with salt, considerable crystallization can be obtained before evaporation begins by introducing into hot water as much of the

substance as it can take up, and then allowing the solution to cool. The formation of frost is really crystallization by deposition from a solution, for the water deposited in solid form was dissolved in the air (§ 96, Part II.). Warm air will hold more water in solution than cold, and, therefore, when warm air becomes cooled, it deposits some of its water in the shape of dew when the cold is moderate, and in the solid form when it is severe. 2. A liquid or a gas may be converted into a solid with a crystalline arrangement. Various examples of this are familiar to you, as the formation of ice from water, the freezing of mercury, and the solidification of melted metals on cooling. The most common example which we have of the conversion of a gas or vapor into a crystalline solid is in the formation of frost from the vapor of water in the air. 3. A mineral substance which is not crystallized may become so by heat, and perhaps some other agencies acting in connection with this. Many rocks, as you will see illustrated in another part of this book, owe their crystalline character to this cause. It is in this way that marble has been made out of common chalk or limestone, it differing from them not in chemical composition, but merely in being crystalline.

20. **Water of Crystallization.**—The crystals of many minerals have water incorporated with them, and, as its presence is essential to their crystalline condition, it is termed their water of crystallization. For a more full statement in regard to this, I refer you to § 164, Part II.

21. **Amorphous and Dimorphous Minerals.**—A mineral is said to be amorphous when it is destitute of all trace of crystalline form. The term comes from two Greek words, *a*, without, and *morphe*, form. When a mineral appears in crystals of two forms, sometimes the one and sometimes the other, it is said to be dimorphous, the first part of the term coming from the Greek word *dis*, twice.

22. **Arrangements of Crystals.**—Crystals forming in groups are arranged variously by the influence of cir-

cumstances. This is exemplified in the huddling together of imperfectly formed crystals in marble, loaf sugar, etc., and in the extremely varied forms of frost-work and other similar crystallizations (§ 16). Some crystals are arranged in the form of a cross, as seen in Fig. 5. Sometimes two similar crystals are united together, and then we are said to have a twin crystal. The crystals of common salt are apt to assume the hopper arrangement represented in Fig. 6. This is because that, as the evaporation takes place, the gravity of the salt makes the mass sink constantly a little below the surface, and each set of crystals is deposited on the upper and outer edge of the preceding set.

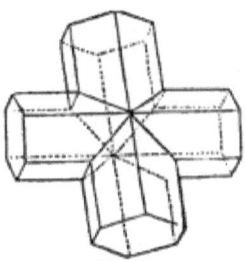

Fig. 5.

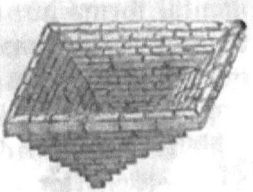

Fig. 6.

23. **Regularity of Form in Rocks.**—We have something akin to crystallization in a rude way in the general lines and faces of rocks. We have the laminated arrangement in the slate rocks, and the magnificent columnar arrangement in the trap rocks, as exemplified in the Giant's Causeway. Then there are joints, so called, running across strata or layers of rock, and dividing them sometimes as evenly as if it were done with a knife. All this will be fully illustrated in the geological portion of this book.

24. **Cleavage.**—Some minerals can have layers chipped or cleaved off, leaving as smooth surfaces as before. Mica is a very familiar example of cleavage from planes. Sometimes the cleavage can be made from angles, as seen in Fig. 7, and sometimes from edges, as seen in Fig. 8. The planes made by cleaving a crystal are

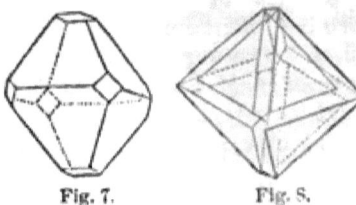

Fig. 7.  Fig. 8.

called its cleavage planes. Minerals sometimes cleave in only one direction, as common mica and foliated gypsum, and sometimes in two, three, or four directions. In many minerals it is difficult to effect cleavage; in some, as quartz, it is impossible, though even in this case it has been done by heating the crystal, and then plunging it into cold water.

**25. Primary Forms.**—While there is vast variety in the forms of crystals as found in nature, mineralogists have discovered by cleavage that the primary or fundamental forms are few in number. There are only thirteen of these forms, which are the solids that have been obtained from the various minerals by cleavage. These are divided into six classes, each class containing those which can be produced by cleavage from each other. The first contains the three solids represented in Figs. 9,

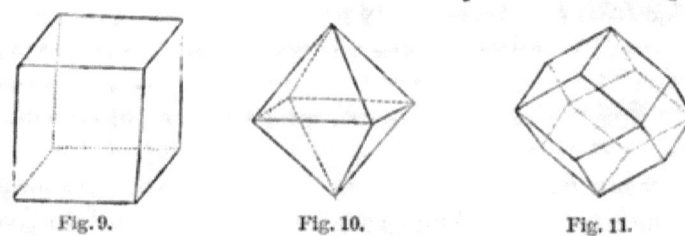

Fig. 9.   Fig. 10.   Fig. 11.

10, and 11, the cube having four equal square sides, the octahedron having for sides eight equilateral triangles, and the rhombic dodecahedron, whose twelve sides are equal rhombic planes.\* I will show by figures how the

---

\* The name dodecahedron is from the Greek words *dodeka*, twelve, and *hedra*, base. A rhombic plane, or rhomb, is a plane with four equal sides, those which are opposite being parallel, and the angles being unequal, two of them being acute and two obtuse. It differs from a square in being oblique-angled instead of right-angled. You
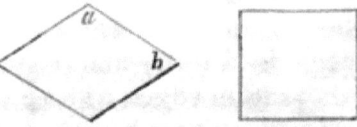
see the difference in the two figures. The angles of the square are all equal, but in the rhomb *a* and its opposite angle are equal, being obtuse, and *b* and its opposite angle are equal, being acute.

cube and the octahedron can be converted into each other by cleavage. If the cube have its angles cleaved as indicated by the dotted lines in Fig. 12, you have as a

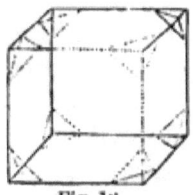
Fig. 12.

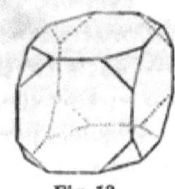
Fig. 13.

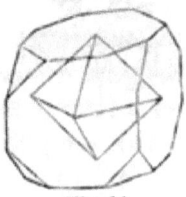
Fig. 14.

result the body represented by Fig. 13. If you continue the process, you will at length obtain the solid represented in the middle of Fig. 14—that is, the octahedron. If,

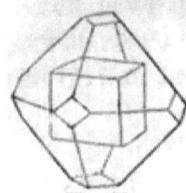
Fig. 15.

on the other hand, you take an octahedron, and cleave it as represented in Fig. 15, you will at length obtain the cube as indicated. These processes, and other similar ones in regard to other forms, may be gone through with very satisfactorily with raw potatoes and a common knife.

26. **Secondary Forms.**—If with a knife you make such cleavages as are represented in Fig. 8, you produce a secondary form. Now, in nature, this result is not produced by cleavage, but by an omission to fill out the whole figure, the omission falling short in various degrees in different cases. The omission may occur on the angles instead of the edges, as seen in Fig. 7.

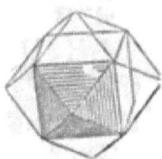
Fig. 16.

Instead of omission there may be addition, as represented in Fig. 16. Here, by an addition in the shape of a low pyramid to each side of a cube, the dodecahedron is produced. The addition on one of the faces of the cube is shaded to make it obvious. The increase is by layer upon layer of particles, each layer having a less number of particles than the preceding one. This is indicated, in a coarse way, by

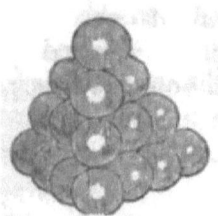

Fig. 17.

Fig. 17, the particles or molecules being in reality so minute that no inequality is perceptible to the eye on the surface of the crystal.

**27. Constancy in the Forms of Crystals.** — Notwithstanding the immense variety in the secondary forms of crystals, there is, in the case of each mineral, a strict adherence to its own form or forms in all essential characteristics. The size and number of faces may differ in different specimens so much as to give them an entirely different appearance, but the mineralogist will discover their relation to each other by observing that the corresponding angles are exactly the same, and by cleavage he will find the same fundamental form. "Crystals are, therefore," says Professor Dana, "the perfect individuals of the mineral kingdom. The mineral quartz has a specific form and structure, as much as a dog or an elm, and is as distinct and unvarying as regards essential characters, although, owing to counteracting causes during formation, these forms are not always assumed. In whatever part of the world crystals of quartz may be obtained, they are fundamentally identical."

**28. Symmetry.** — In all the modifications of crystals seen in the production of secondary forms there is a wonderful symmetry. If one part is modified, all the corresponding parts are modified, and in the same way, either over the whole crystal or over just one half of it. Thus, if we find a plane in place of an edge, we shall find the other edges also replaced by planes of precisely the same width, as seen in Fig. 8; and so of the angles also, as represented in Fig. 7. So, if there be two planes in place of any edge or angle, there will be found the same planes replacing the other edges or angles; and the same is true if there be many replacing planes, as there often are. You see how this secures a regularity pleasing to the eye in the midst of extreme variety, thus adding largely to the beauty of the mineral world. Observe that there is here some analogy to the symmetry that appears in living forms. The two halves of a face, for example, are generally exactly alike, just as are the corresponding parts of a crystal. What the agency is that produces the result is in both cases a wonderful mystery. All that we know is that in the one case the deposition of matter is guided unerringly by a living agency, and in the other by one that is not living.

## CHAPTER III.

### CARBON AND ITS COMPOUNDS.

I PROCEED now to treat of different kinds of minerals, grouping them according to their natural affinities.

29. **Diffusion of Carbon.**—As carbon is one of the four grand elements in the composition of vegetable and animal substances, there is a constant interchange in regard to it between the vegetable, animal, and mineral worlds, and therefore it is widely diffused, and appears in various combinations. Combined with oxygen, it exists every where in the atmosphere in the form of carbonic acid, which is one of the gases that are more or less mingled or dissolved in the water of the earth. The carbonates are important salts, the most important being the carbonate of lime, which appears in the various forms of chalk, limestone, and marble, and in the animal world makes the hard covering of shell-fishes and the skeletons of the coral animals. The immense stores of coal laid up in the bowels of the earth is almost wholly carbon.

30. **The Diamond.**—In strong contrast with the immense provision of carbon in the form of coal, this element is laid up here and there in very small quantities, equally for the use of man, in the most costly and splendid of gems. That the diamond is pure carbon is proved

by burning it in oxygen gas, the product being carbonic acid gas, a compound of carbon and oxygen, just as in the burning of charcoal. As in pure charcoal and anthracite we have carbon alone, the diamond differs from them merely in the arrangement of the atoms. It is this alone that causes the immense difference in characteristics. When the diamond is colored, as it sometimes is, there is coloring matter intimately mingled with the carbon, as glass is colored with something added to it. But the quantity of coloring matter is so exceedingly minute that it could not be detected unless a considerable amount of diamond were burned, an experiment so expensive that I presume that it never has been tried.

31. **Qualities of the Diamond.**—It is the hardest of all substances. Its lustre is peculiar; and other gems that are similar to it in this respect are said to have an adamantine lustre. Its reflection of light is exceedingly brilliant. It is commonly colorless, but sometimes red, yellowish, orange, green, or black. The rose diamond is highly valued for the beauty of its color, and so also is the green diamond. The black diamond is very rare, and therefore commands a high price, although it has no beauty.

32. **Size of Diamonds.**—In speaking of the size of diamonds the term *carat* is used. This is the name of a bean, which was used in its dried state by the natives of Africa in weighing gold, and in India in weighing diamonds. Though the bean is not used for this purpose now, the name is retained, and the carat is nearly four grains Troy. The largest diamond known is in the possession of the Great Mogul. It weighed originally 900 carats, or 2769 grains, but it was reduced by cutting to 861 grains. It is of the size of half a hen's egg, and is in that form. The Pitt or Regent diamond weighs 419 grains. The famous Koh-i-noor weighed originally 186 carats, but was reduced by recutting one third.

33. **Cost of Diamonds.**—The cost of diamonds depends

upon the weight, the purity, and the color. A diamond which, after being cut, weighs one carat, is worth commonly £8. The price increases largely with the size, for one weighing four carats would be worth £128, and one of ten carats would be £800. When we get up to twenty carats the prices rise much more rapidly. The Regent diamond is estimated at £125,000.

**34. Cutting of Diamonds.**—The art of cutting and polishing diamonds was unknown till 1456, when it was discovered by Louis Berquen, of Bruges. The process is thus described by Professor Dana. "The diamond is cut by taking advantage of its cleavage, and also by abrasion with its own powder, and by friction with another diamond. The flaws are first removed by cleaving it, or else by sawing it with an iron wire, which is covered with diamond powder—a tedious process, as the wire is generally cut through after drawing it across five or six times. After the portion containing flaws has thus been cut off, the crystal is fixed to the end of a stick, in a strong cement, leaving the part projecting which is to be cut; and another being prepared in the same manner, the two are rubbed together till a facet is produced. By changing the position, other facets are added in succession till the required form is obtained. A circular plate of soft iron is then charged with the powder produced by the abrasion, and this, by its revolution, finally polishes the stone. To complete a single facet often requires several hours." The expense of cutting the Regent diamond was estimated at £5000 sterling, and the mere filings at £7000 sterling.

**35. Uses of the Diamond.**—The most familiar use of the diamond is cutting glass. It is also used for lenses in microscopes. Diamonds that can not be worked are sold under the name of *bort*, for various uses. Splinters of bort are made into fine drills for drilling artificial teeth and gems of various kinds.

**36. Localities of Diamonds.**—Diamonds are found in

various parts of India and of Brazil, in Africa, in the island of Borneo, and in the Urals of Russia. The original rock of the diamond appears to be a quartz rock; but the diamonds are almost always obtained from amid the sands and pebbles which have come from the rocks, and are scattered in the rivers and brooks. In Brazil, collections are made of these sands and pebbles, and by washing them in a series of boxes the diamonds are discovered and gathered. If a negro be so fortunate as to find a diamond weighing $17\frac{1}{2}$ carats, he gains a boon which is above the price of gems—the boon of liberty.

37. **Graphite.**—Graphite, or Plumbago, commonly called black lead, although there is not a particle of lead in it, is composed of carbon, with a small proportion of iron, commonly to the amount of from 4 to 10 per cent. Some have supposed it to be a carburet of iron; but it is not a chemical compound. It is only a mixture of the carbon and iron, and that the iron is not essential is shown by the fact that in some cases there is scarcely a trace of it. Graphite is soft, and has a shining lustre. It is sometimes compact, and sometimes crystalline, usually in the foliated form, but occasionally in six-sided prisms. It is used in making lead-pencils, in lessening the friction of machinery, in giving a gloss to stoves, etc. Commonly, in preparing graphite for pencils, the solid mineral is cut up into pieces of the requisite size; but a method has of late been adopted by which the mineral is finely powdered, and then, by great pressure, made into solid sheets, from which the pieces are obtained.

38. **Coal.**—The varieties of coal are divided into two classes, the bituminous and the non-bituminous. The former have, in addition to the carbon, hydrogen, which in the combination produces carbureted hydrogen, or common illuminating gas, and it is the burning of this, that is, its union with the oxygen gas of the air, that causes the flame. In the non-bituminous, the anthracite, on the other hand, we have only the blue flame in the

beginning of its combustion, which comes from the carbonic oxyd gas as it unites with the oxygen of the air to form carbonic acid. The formation of this gas in the anthracite fire is fully explained in § 67, Part II. It is from the bituminous coal that the illuminating gas is produced, there being no hydrogen in the anthracite for its formation. The explanation of the process is this: By the application of heat the hydrogen is driven off, and it takes with it enough of the carbon to make it carbureted hydrogen; but most of the carbon is left behind in a light, porous condition, to which is given the name of *coke*.

39. **Anthracite.**—This is the common name given to coal which is either mostly or wholly free from bitumen. It is supposed that it was subjected to more heat than the bituminous was after its formation, by which the bituminous elements were driven off. The coke, which is left after the making of illuminating gas, is very much the same thing as anthracite, except that, as it has not been subjected to great pressure, it has not the same closeness of structure. Anthracite is hard, and so is often called very appropriately stone coal. From 80 to 90 per cent. of it is carbon. Then there is from 4 to 7 per cent. of water, and there are some impurities, as silica, alumina, lime, magnesia, etc. The redness of the ash of some kinds of coal comes from the presence of oxyd of iron. The *slag*, or glassy substance which you find in the refuse of the combustion, is composed of silicates. These do not form with the ordinary burning of coal in grates, because the heat is not intense enough. The clinker which collects on the inside of stoves is slag. The reason that oyster-shells put into the fire clear off the clinker is that the lime makes the silicates more fusible, and so dislodges the clinker, mingling it with the burning coal. Anthracite has a fine lustre, and often an iridescent play of colors. It is capable of a high polish, and has sometimes been made into inkstands and other articles.

40. **Bituminous Coal.**—This is softer than anthracite,

and has less lustre. Its chief varieties are the *pitching* or *caking* coal, which in burning is apt to cake or unite into solid masses, and the *cannel* coal, which burns with so clear a flame that it has sometimes been used as candles. Inkstands, boxes, etc., are made from it. The mineral called jet, so much used in jewelry, is allied to cannel coal, but has a much deeper color, and is capable of a brilliant polish. Brown coal, or lignite, is not so perfectly formed coal as the other varieties, and the structure of the original wood is often apparent in it.

**41. Coal of Vegetable Origin.**—That coal is made from plants might be properly inferred from the remains and impressions of the plants found in the layers of rocks between which the coal is packed, and sometimes found even in the coal itself. But the proof is more positive than this. Vegetable structure is found in the coal by microscopical examination. A piece of anthracite coal which has been partially burnt is used, because the vegetable cells, being somewhat silicious, remain unaltered in the burning out of a portion of the carbon. In Fig. 18, *a*, we have represented a small bit of such a piece of anthracite as seen through the microscope. The ducts lie side by side, as they did ages ago in the woody fibre of the plant. In *b* two of the ducts are very highly magnified, the white spaces showing us the silica, and the black lines the carbon which was not burned out. The evidence from such an examination is not as clear in soft bituminous coal, because the original texture is not so well preserved.

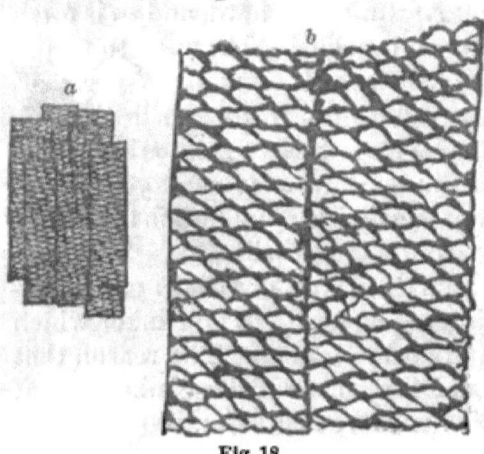

Fig. 18.

**42. Why Coal is called a Mineral.**—Coal being thus of vegetable origin, and retaining the evidences of this in its very texture, it might at first thought appear that it could hardly be called a mineral. The reason that it is proper to call it so I will point out. The carbon, which is the chief component of all kinds of coal—almost the only component of one kind, the anthracite—does not exist in the vegetable as carbon, but, chemically united with oxygen and hydrogen, it forms woody fibre. It is just as sulphur does not exist as sulphur in sulphate of lime (gypsum), but as a part of a compound united chemically in it with oxygen and lime, as carbon in wood is united chemically with oxygen and hydrogen. Now when wood is made to produce coal a *chemical* change occurs, a real decomposition is effected, so that carbon now appears in its uncombined state. The form, indeed, of woody fibre may remain, but it is no longer wood—the compound of carbon, oxygen, and hydrogen. These elements are separated from each other; and if they are combined, as may be the case in bituminous coal, it is a recombination after separation. A farther chemical change or decomposition is produced when the coal is burned, all the carbon then passing off by forming carbonic acid gas with the oxygen of the air. The first process is as strictly and fully a chemical change as the last, and it is by chemical changes that minerals are produced. We have here a clear case of the passage of an organized substance under the control of living agencies over into the mineral kingdom, where chemical laws bear sway.

**43. Peat.**—In peat we have an example of imperfect formation of coal, the approach to the completion of the process being different in different cases. It accumulates in swampy places, mostly from the growth of mosses of one genus, *Sphagnum*. The roots continually die below as the plant grows upward; and as these roots, by the partial decomposition under water, change into peat, a bed of great thickness may after a time be formed.

There is sometimes a depth of forty feet in a peat-bed. The beds in some countries are very extensive. They cover one tenth of Ireland, and one of them is fifty miles long and two or three in breadth. It is estimated that the amount of peat in Massachusetts is 120 millions of cords.

**44. Diffusion of Coal in the Earth.**—The stores of coal in the earth are immense. The provision which has been made by the Creator in this respect for the wants of man, and the means which have been adopted for bringing the coal within his reach, are among the highest evidences of far-reaching design and system in the construction of the earth. This point, barely alluded to here, will be fully developed in another part of this book.

**45. Amber.**—This is a resin from some tree which has been chemically changed, or mineralized, as it is expressed. It is found in lumps of various sizes, from small bits up to even the size of a man's head. It is commonly of a yellowish color, has considerable lustre, and is susceptible of a polish. It is appropriated to ornamental purposes, but it is not highly prized in this respect, as it is soft, yielding readily to the knife, and, besides, can be easily counterfeited. It is used in making a superior transparent varnish. There are sometimes seen in specimens of amber insects and parts of insects, which became enveloped in the resin in its semi-liquid state. Such specimens are highly prized, because it is supposed these insects lived ages before the creation of man. Imitations, therefore, have sometimes been made, and have been palmed off upon those who are not able to distinguish insects of those early ages from the insects of the present day.* Amber becomes electric when it is rubbed,

---

* In the geological part of this book you will learn that both the animals and plants of the early ages of the earth differed, in some respects, from those which exist at the present time. The imposture referred to reminds me of one which was once detected in the court-room when a will case was on trial. The will, on being held up to

and therefore its Greek name *electron* gave rise to the term electricity.

46. **Bitumen.**—There are various kinds of bitumen, some solid and some fluid. They are supposed to be produced from vegetable matters buried in the earth; and as they are commonly found in the neighborhood of active or extinct volcanoes, it is inferred that they have been driven to the surface by the internal heat of the earth. I will notice some of the varieties. *Asphaltum*, or mineral pitch, is solid. It was the chief ingredient in the cement used in building the walls of Babylon and of the Temple in Jerusalem. It has lately been successfully employed in forming a composition for paving streets, and a cement for covering roofs. It is very abundant on the shores of the Dead Sea, which is called Asphaltites for this reason. The most remarkable locality of it is the Pitch Lake in the island of Trinidad. It is about a mile and a half in circumference, and while the asphaltum is near the shores sufficiently hard at most seasons to sustain men and quadrupeds, it grows soft and warm as you go toward the centre, and there it is in a boiling state. *Petroleum* is a dark fluid, which becomes solid on exposure to the air. This is so abundant in the Burmese Empire that it is used as lamp-oil, and as fuel, by being mixed with ashes or earth. The most powerful springs of it there are on the Irawady. In one locality there are 520 wells, yielding annually 400,000 hogsheads. There are many localities of petroleum in this country. It was formerly collected and sold by the Indians; and, as the Senecas were prominent in the trade,

the light, was found to have been written on paper which, by its water-mark, was shown to have been manufactured subsequent to the date of the document. So, in these imitations of amber, the date fixed by the recent character of the insects shows, to the practiced eye of one who understands the distinctions between insects of the present age and those of ages previous to the creation of man, that the specimens are not real amber.

it acquired the name of Seneca oil, which it still retains. *Naphtha* is a light, limpid fluid of a yellowish color, and is a purer article than petroleum, from which it can be obtained by distillation. It is a hydrocarbon; that is, it is composed of hydrogen and carbon. As there is no oxygen in its composition, it is used, as you learned in Part II., by the chemist to preserve potassium and sodium in their metallic state.

47. **Carbonic Acid.**—The qualities of this gas you learned in Part II., Chap. III. It is constantly furnished to the air from the respiration of animals, from fires, and from the decomposition of vegetable and animal matters. As it is constantly absorbed by the leaves of plants, its undue accumulation in the atmosphere is prevented. There are localities where this gas is produced in the earth in large quantities. In some mineral springs, as at Saratoga, it gives briskness and a slight pungent taste to the waters. This large production of it appears in the neighborhood of some volcanoes. A cavern near Naples, the Grotto del Cane, has a world-wide reputation on account of the abundance of this gas in it, which fills the cavern up to the level of the lower margin of its entrance. This gas has a great agency in relation to one of the most important and extensive of the rocks of the earth, the limestone, which is a combination of this gas with lime. The carbonate of lime of the rocks is continually being dissolved in the water that comes in contact with them; and then the shell-fish and the coral animals appropriate this to make their shells and skeletons, which eventually become, as you have seen (§ 8), limestone rock, thus returning the carbonate of lime to the source from which it came. Now it is the carbonic acid in the water which enables it to dissolve sufficient of the limestone to supply the shell-fish and the corals with this material. For a full elucidation of this interesting point I refer you to Part II., § 306.

48. **Carbureted Hydrogen.**—This, which is the common

illuminating gas, produced from bituminous coal artificially, is produced naturally in many localities, and probably comes from the action of heat upon bituminous coal, or some other substance of a kindred character. In the village of Fredonia, New York, near Lake Erie, this gas issues so abundantly from a slate rock that it is collected, and is used by the inhabitants for lighting the place.

## CHAPTER IV.

### SULPHUR AND ITS COMPOUNDS.

**49. Native Sulphur.**—This substance is of a pale yellow color. It is found both crystallized and uncrystallized. Its crystallized form is an octahedron, as seen in Fig. 19. When sulphur is melted, and then left to crystallize, the crystal takes the form seen in Fig. 20, which is very different from the forms which are assumed in nature. Sulphur is often found in fissures and cavities in lava. The most important locality of sulphur is Sicily. Sixteen or seventeen thousand tons are imported annually into England from Sicily, to say nothing of the supply to other countries. Much of the sulphur of commerce is obtained from the sulphurets of iron and copper. The sulphur is driven off from these ores by heat, and is collected in brick chambers. It is one of the ingredients of gunpowder, and is used in bleaching and in the manufacture of sulphuric acid, as described in § 249, Part II.

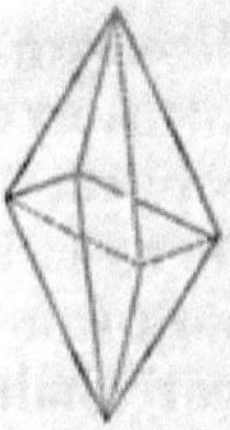

Fig. 19.
Fig. 20. 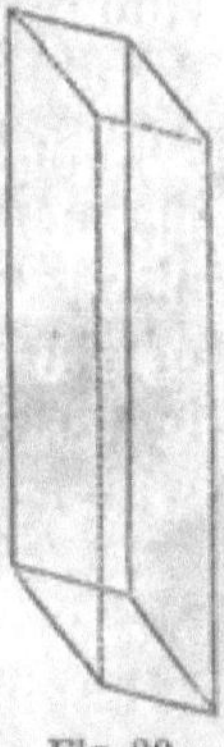

**50. Sulphurets of Iron.**—The bisulphuret of iron is one of the most common ores in the earth, and has been known from ancient times, when it received the name of pyrites, from the Greek word *pur*, fire, because, as Pliny says, "there is much fire in it," referring to its readily striking fire with steel. It occurs, when crystallized, in the primary forms of cube, octahedron, and dodecahedron (Figs. 9, 10, and 11), and in a variety of secondary forms. Its color is a bronze-yellow, and the lustre is often brilliant. As the lustre is metallic, it has been often supposed by common people to be gold, and specimens which have been found are every now and then taken to some person of known scientific character in the community for inquiry on this point. It has therefore received the name of "fool's gold." The distinction between it and gold is, however, easily made, for the pyrites is very hard, is not malleable like gold, and on the application of strong heat it gives off sulphurous fumes. It is not much prized as an ore for obtaining iron, because of the difficulty of ridding it entirely of the sulphur; but it is much used for obtaining sulphate of iron (green vitriol or copperas), sulphuric acid (oil of vitriol), and sulphur. There is a magnetic iron pyrites, which is softer and liable to tarnish. There is also an arsenical iron pyrites, nearly half of which is arsenic. Its color is silver-white, with a shining lustre, and on being struck with steel it gives a spark with a garlic odor.

51. **Copper Pyrites.**—This is not a sulphuret of copper, but of copper and iron, nearly one third of it being iron. It has a deeper color than iron pyrites, is softer, yielding readily to the knife, and does not strike fire with steel. Its color is such that it looks like gold, but it is at once distinguished from it by its crumbling under the knife. Much of the metal copper is obtained from this ore, and sulphate of copper (blue vitriol) is largely manufactured from it. There is a copper ore called copper glance, a sulphuret in which the proportion of iron is very small. It is of a lead-gray color. There is also a sulphuret of a dark steel-gray color, of a very compound character, there being in its composition six ingredients—sulphur, copper, antimony, arsenic, iron, zinc, and silver. Sometimes the

silver in it amounts to 30 per cent., and then it is called *argentiferous gray copper ore*.

52. **Sulphuret of Lead.**—The common name of this mineral is *galena*. Its crystals are in the form of the cube and its secondaries. Its color is lead-gray, and its lustre is shining. There is often sulphuret of silver incorporated with it, and it is then called *argentiferous galena*. It is the ore from which the metal lead is obtained, native lead being exceedingly rare. I have told you how the galena is freed from the sulphur in Part II., § 193, and in § 210 how the silver is obtained from the galena when it is argentiferous. There are vast quantities of galena in various parts of the United States, especially in Missouri, Illinois, Iowa, and Wisconsin. It is so abundant in the last-named state that the miners seldom take the trouble to get out the ore deeper than twenty-five or thirty feet. Three millions of pounds have been obtained from one spot which is not over fifty yards square.

53. **Sulphuret of Silver.**—This mineral, commonly called silver glance, occurs in the form of the dodecahedron and its secondaries. Its color is lead-gray, and it has a metallic lustre. There is also a sulphuret of silver and antimony, which is of a black color, and therefore is sometimes called black silver. There is also a sulphuret of silver and iron, and one of silver and copper.

54. **Sulphuret of Antimony.**—This sulphuret appears in prisms, and has a gray color and metallic lustre very much like those of galena. It is the ore from which is obtained nearly all the antimony of commerce. There are also some sulphurets of lead and antimony combined together.

55. **Sulphurets of Arsenic.**—There are two sulphurets of arsenic. One is of a red color, is translucent, sometimes transparent, and its crystals are usually in the prismatic form. It is called *realgar*. The other, called *orpiment*, is of a lemon-yellow color. Both are used as pigments, and also as coloring substances in pyrotechnic mixtures. They are obtained chiefly in Koordistan and China.

**56. Sulphuret of Mercury.**—This is more abundant than any of the other ores of mercury, and most of this metal is obtained from it. It is called both cinnabar and vermilion. The latter is its name as sold in the shops. It is used as a pigment, and is the coloring matter in red sealing-wax. The principal mines of this ore are in Austria, Spain, Peru, and Upper California.

**57. Sulphuret of Zinc.**—This mineral, commonly called blende, the *Black Jack* of miners, appears in octahedrons, dodecahedrons, and forms that are secondary to these. It has a waxy lustre, and when a cleavage face is made the lustre is brilliant. It is very apt to be found in company with lead ores, and it abounds in the lead mines of Missouri and Wisconsin.

**58. The Three Vitriols.**—The green vitriol (sulphate of iron), the blue (sulphate of copper), and the white (sulphate of zinc), are all familiar to you. They exist in nature in company with the sulphurets of the same metals, from which they are really produced. A sulphate differs from a sulphuret in having oxygen in it, as you learned in Part II. Now if a sulphuret, of iron for example, be exposed to the air in a moist state, it is decomposed, and the sulphur and the iron, absorbing oxygen from the air, become, the one sulphuric acid and the other oxyd of iron, and these two uniting form sulphate of iron. So also sulphate of copper is formed from sulphuret of copper, and sulphate of zinc from sulphuret of zinc.

Green vitriol, or copperas, is largely used in dyeing and tanning, because, in connection with an ingredient in nut-galls and many kinds of bark, it gives a black color. Common ink is essentially a combination of copperas with this ingredient.

**59. Sulphate of Lead.**—This is usually in company with galena, and is the result of its decomposition, after the manner indicated in speaking of the vitriols. Its color is white, or light gray, or **green**. Its crystals are sometimes splendid.

**60. Gypsum.**—This is sulphate of lime. When crystallized and free from all impurities it has a pearly lustre, and the clearness of glass. About a fifth part of it is water. This water is driven off in preparing it for making casts, ornamental work for walls, etc. The manner of using it for these purposes is described in § 320, Part II. Gypsum is a white and soft mineral, and appears in many forms, some of which are very beautiful. One of these is the *satin spar*, so called from the splendid lustre of its delicate fibrous arrangement. Another, called alabaster, which is generally snowy white, being compact, with a fine grain, is cut into vases and ornaments of various kinds. Sometimes gypsum is composed of exceedingly thin leaves, laid together so evenly that a multitude of them make a crystal clearer than the clearest glass. The name selenite, which has been given to this and some other varieties of gypsum, comes from *selene*, the Greek word for moon. There is an anhydrous sulphate of lime, that is, one which has no water incorporated with it, the term coming from two Greek words, *an*, without, and *udor*, water. Gypsum occurs abundantly in many parts of this country. Fine specimens are found near Lockport, New York. As before noticed (§ 16), in the Mammoth Cave of Kentucky alabaster appears with its crystals arranged in various forms of flowers, branches of shrubbery, vines, etc.

Common limestone has sometimes been mistaken for gypsum by persons who are ignorant of chemistry and mineralogy. Professor Hitchcock relates a case of this kind. A farmer supposed that he had found gypsum on his farm, and his neighbors, believing it, bought large quantities of the material for agricultural purposes. After grinding up a considerable amount of it, some one accidentally discovered that it was limestone. The error might have been avoided by simply testing the substance with acid. A drop would have occasioned an effervescence, showing that it was carbonate, and not sulphate of lime, the acid taking the lime to itself, and setting free the carbonic acid gas. See § 60, Part II.

61. **Sulphates of Magnesia and Soda.**—The slender, spicula-like crystals of Epsom salts (sulphate of magnesia), which are probably familiar to you, are prisms. Minute crystals of this mineral are often present in the earth on the floors of the limestone caves in the western part of this country. In the Mammoth Cave of Kentucky feathery masses of the crystals adhere to the roof, looking like snow-balls. The common name of this mineral came from the fact that at Epsom, in England, the waters of the springs hold it in solution. The crystals of Glauber's salt are also prisms, but they are coarse compared with those of Epsom salts. Both of these salts are present in sea-water.

62. **Sulphate of Baryta.**—This is so heavy a mineral that it is called heavy spar. The crystals, which are translucent, sometimes transparent, are often very beautiful. It is found abundantly in some localities in this country, as Cheshire, Conn.; Pillar Point, N. Y.; near Fredericksburg, Va. It is extensively ground up for use in paints, being mixed with white lead. The mixture has various names, according to the proportions of the various ingredients: *Venice White*, when there are equal parts of the two; *Hamburg White*, when the lead is half the weight of the sulphate of baryta; and *Dutch White*, when it is one third.

63. **Sulphuric and Sulphurous Acids.**—That intense acid, sulphuric acid, or oil of vitriol, as it is commonly called, is occasionally found near volcanoes and sulphur springs. Sulphurous acid, the pungent and suffocating gas which is produced whenever sulphur is burned, is often very abundant about volcanoes when they are in action.

64. **Sulphureted Hydrogen.**—This very offensive gas is common at sulphur springs, and it is by its agency that articles of silver are so readily blackened in such locali-

ties, the sulphur of the gas uniting with the silver to form a sulphuret of that metal. This gas is also sometimes generated about volcanoes.

## CHAPTER V.
### METALS AND THEIR ORES.

**65. Native Metals.**—A metal is said to occur native when it is found either pure or mingled with some other metal in the form of an alloy; in other words, when it is not united *chemically* with any other substance.

**66. Ores.**—When a metal is united chemically with any substance, the compound is called an ore. The metal is in this case said to be mineralized. The most common of these compounds are sulphurets, oxyds, and carbonates. The word ore is sometimes used in a less strict sense for alloys, and even native metals are often termed ores. On the other hand, it is proper to state that to the compounds of some of the metals the term is not applied at all. I refer to the earths and alkalies, which are combinations of certain metals with oxygen, to their salts, and to the combinations of these same metals with chlorine, iodine, etc. These metals, potassium, sodium, etc., have been discovered comparatively at a recent date, are never found native, and, when obtained by chemical processes, are, most of them, preserved in their metallic state with difficulty. You see, then, the propriety of not calling their compounds ores. It is to the compounds of the well-known metals only that the term is applied.

**67. Positions and Associations of Ores.**—The ores of metals are sometimes scattered here and there in rocks, in collections small and large; but commonly they are in veins, or *lodes*, as they are called, or in layers between layers of rock. They are associated with quartz, carbonate of lime, and various other minerals. Often two or more different ores are mingled together.

**68. Obtaining Metals from their Ores.**—Sometimes this is a very simple process. For example, in the case of bismuth, all that is necessary is to heat the pounded ore, and the melted metal runs out. So, if you have a sulphuret of lead, or mercury, or antimony, heat will fully decompose the compound, driving off the sulphur. But sometimes the use of some other substance is required for the decomposition of the ore. Thus, if we have an oxyd of iron, we heat it with charcoal. Here the oxygen quits the iron to unite with the carbon, forming carbonic acid, which flies off, leaving the metallic iron.

**69. Gangues and Impurities.**—The rock in which an ore is found is called the *gangue*. Much of this is separated from the ore in collecting it, and much more of it, perhaps, by the process called *washing*, in which the material, coarsely powdered, is subjected to a current of water, that washes away the lighter pieces, leaving those which are rendered heavy by the presence of the metal. The impurities which are often mingled with ores are got rid of in various ways. Sometimes *fluxes*, so called, are used for this purpose. This is done with most iron ores. There is commonly mixed with these ores quartz or clay; and as quartz is pure silica, and silica constitutes 75 per cent. of clay, the ore is reduced by strongly heating it together with a substance which will form a silicate, that is, a glass, with the silica (§ 342, Part II.). Such a substance is common limestone.

Other processes will be noticed hereafter in the case of particular ores.

**70. Iron.**—The most common of the ores of this metal are oxyds and sulphurets. Of the latter I have already spoken in § 50. Its ores are more abundant in the earth than those of any other metal, because it is the most extensively and variously useful of all the metals. They are the common coloring ingredients of rocks, and therefore of soils. Red and yellow are the most frequent colors, but they color also a dull green, brown, and black.

Iron is present in a small amount in many vegetable and animal substances, and in our blood it is an essential, though a minute ingredient (§ 663, Part II.).

**71. Meteoric Iron.**—Native iron has not been ascertained to occur except in meteorites, and there it is alloyed with nickel, and with a small amount of other metals—tin, cobalt, copper, and manganese. The mass in the cabinet of Yale College, from Texas, weighing 1635 pounds, contains from 90 to 92 per cent. of iron, and from 8 to 10 per cent. of nickel, the mixture of the two metals not being uniform throughout the mass. There is one mass of meteoric iron in South America which is supposed to weigh 30,000 pounds.

**72. Magnetic Iron Ore.**—This ore has this name from its magnetic properties. It is an oxyd of iron, of an iron-black color, generally in granular masses, but sometimes in distinct crystals, which are octahedrons or dodecahedrons, or their secondary forms. This very superior ore of iron is as widely disseminated as any ore of this metal. Nearly all the Swedish iron ore is of this kind. In Sweden and Lapland there are mountains of it. When a mass of this ore is in a state of magnetic polarity it is a *lodestone* or *natural magnet*. The lodestone was first found in the province of Magnesia, and was called *magnes* by Pliny, and hence the terms magnet and magnetism.

**73. Hematite.**—This, like the magnetic iron ore, is an oxyd, but is distinguished from it by its powder being red. It is this which has given it its name, from the Greek word *haima*, blood. This mineral appears in various forms, generally in granular mass, in laminæ, or earthy, and easily powdered. What is called red chalk is one of the varieties. This ore is abundant in this country. The two iron mountains in Missouri consist for the most part of this ore piled up "in masses of all sizes, from a pigeon's egg to a middle-sized church."

**74. Limonite, or Brown Iron Ore.**—This is not merely an oxyd of iron, but a *hydrated* oxyd, containing water to about 14 per cent. One of its forms is yellow ochre, which is used as a common material in paint. The powder of this ore is also used for polishing metallic surfaces.

**75. Chromic Iron.**—This mineral is composed chiefly of the oxyds of two metals, iron and chromium. The oxyd of the latter acts as an acid, as mentioned in § 301, Part II., and the mineral is therefore said to be a chromate of iron. There are, however, two other components, alumina and magnesia, which vary in quantity in different specimens. This mineral is quite abundant at Barehills, near Baltimore, in Lancaster county, Pennsylvania, and several other places in this country. Its crystals are octahedrons. It is valuable in the manufacture of the chrome pigments, of which the chrome yellow is the principal.

Fig. 21.

**76. Carbonate of Iron.**—This mineral varies in color from light gray to dark brown or nearly black on exposure. Its crystals sometimes have curved faces, as represented in Fig. 21. Metallic iron is extensively obtained from this ore.

**77. Native Copper.**—This occurs in company with ores of the metal, commonly in the neighborhood of igneous rocks—that is, rocks which have been made and thrust upward in the crust of the earth by the agency of heat. There is often silver with the copper, either intimately mixed with it, making an alloy, or collected by itself in small masses or in strings. Copper is next to iron in abundance. There are famous mines in Cornwall, England, in Brazil, and in Siberia. Perhaps the most extraordinary copper region in the world is in the vicinity of Lake Superior. It is found there in veins filling up fissures in the rocks, and it is cut out in monstrous blocks with chisels and drills. One mass, weighing nearly 4000 pounds, was carried to Washington, and a large mass has been since got out which was estimated to weigh 200 tons. As you will see in another part of this book, the geologist has found clear proof that the native copper of this region was produced in Nature's great furnaces from ores of the metal ages upon ages before the creation of man.

The sulphurets of copper and the sulphate were noticed in the preceding chapter. I go on now to notice the principal of the other ores of this metal.

78. **Oxyds of Copper.**—There are two oxyds, a red and a black one. The metal can be obtained from them by heating with charcoal, the oxygen uniting with the carbon to form carbonic acid gas, which flies off.

79. **Carbonates of Copper.**—There are two carbonates. One of them, called malachite, is of a light *green* color, and, as it is capable of a high polish, it is used in various ways for ornamental articles. Its value, when manufactured as veneering or inlaid work, is about three guineas per pound, and there are at least two pounds and a half in a square foot of finished work. In Russia, where large pieces of it can be obtained, slabs for tables, mantle-pieces, vases, etc., are made from it. The *blue* carbonate sometimes presents splendid crystals.

80. **Silicate of Copper.**—This mineral, a compound of silica or silicic acid with oxyd of copper, has a bluish-green color. Sometimes the green carbonate and this silicate are united in one mineral.

81. **Lead.**—Native lead is exceedingly rare. Its most common ore, galena, and the sulphate I noticed in the preceding chapter. The *oxyd* of lead, minium, is what is commonly called *red lead*. The carbonate is the *white lead* of commerce. The *chromate* of lead is the chrome yellow used by painters.

82. **Tin.**—The native metal is exceedingly rare. The chief ore is an oxyd, the sulphuret being seldom found. This is an ingredient in pewter and bronze, and in the amalgam put on the backs of mirrors, and some of its salts are employed in dyeing. The tin used in making

tin-ware is iron in the form of sheets covered with tin, which does not oxydize as readily as tin does. The chief tin mines are at Cornwall, in England, in the island of Banca, in Malacca, and in Austria. It is supposed, from some allusion in ancient history, that the Cornwall mines were worked some centuries before the Christian era.

83. **Zinc.**—This metal never occurs native. I have already noticed its chief ore, the sulphuret, in § 57, and the sulphate in § 58. The other ores which are at all prominent are the red oxyd and the carbonate. For the uses of this metal I refer you to § 198, Part II.

84. **Antimony.**—This metal is occasionally found native. Its usual combinations are with sulphur, or with sulphur and lead together. It is sometimes also combined with oxygen, arsenic, lime, nickel, silver, and copper. The chief use of the metal is in the composition of type-metal. In medicine there is used a double salt, in which antimony is one of the bases, the tartrate of antimony and potash.

85. **Cobalt.**—There are two arsenical ores of this metal—that is, ores in which it is combined with arsenic. *Zaffre* is a beautiful coloring material prepared from the ores of cobalt. *Smalt* is glass colored with zaffre and reduced to a fine powder. This is used to give a delicate blue tinge to writing paper and to linen.

86. **Nickel.**—Native nickel is never of terrestrial origin, but it is one of the ingredients in the alloy which we have in meteorites, as stated in § 71. It has two arsenical ores, a carbonate, various compound sulphurets, etc. German silver, so called, is an alloy of copper, nickel, and zinc.

87. **Bismuth.**—This is found native, and also combined with oxygen, carbonic acid, silica, and the metal tellurium. It is an ingredient in the most superior kind of type-metal, in the "mosaic gold," and in an alloy used in soldering pewter. Most of it comes from one locality, Schneeberg, in Saxony. It is obtained from the rocks

in which it is present by powdering coarsely, and then exposing to strong heat in a kiln. The bismuth is melted only, and runs into a trough at the bottom.

88. **Manganese.**—The ores of this metal are some of them composed of but two ingredients, as the oxyds and sulphurets; and others are very compound, containing silica, iron, lime, magnesia, etc. The peroxyd is used by the chemist for obtaining oxygen gas, and is employed largely in bleaching. Some of the salts of this metal are used in calico printing. Manganese gives a violet color to glass.

89. **Mercury.**—This is the only metal that is liquid in all ordinary temperatures. It becomes solid at 39° below zero, and therefore can not be used in thermometers in the arctic regions. It crystallizes in cubes. It is sometimes found native, commonly in globules scattered here and there, but sometimes in such quantities that it can be dipped up in pails. It is related that the mines of mercury in Mexico were discovered by a hunter, who, as he clambered up a mountain, caught hold of a shrub, which, giving way at the root, let out a stream of what he supposed was liquid silver. The rapidity with which this metal runs, occasioned by its great weight, gave to it the name of quicksilver. The purity of the mercury is judged of by the workmen by watching it as it is dropped upon glass. If in running, instead of preserving fully its characteristic globular shape, it forms a queue, or drags a tail, as it is expressed, there is lead or some other impurity present. Mercury is used in a variety of ways. It is the liquid used in barometers, and generally in thermometers. It is used for silvering mirrors, an amalgam being formed with tin for this purpose. Its disposition to amalgamate with other metals is made use of extensively in extracting gold and silver from the impurities with which they are mixed, as described in § 213, Part II. Then there are various preparations and combinations of this metal employed in medicine. The

most common ore of mercury, the sulphuret, was noticed in § 56.

90. **Silver.**—This metal is found pure, in alloy, and in chemical combination with sulphur, arsenic, chlorine, various acids, etc. When it is found native it is not usually alone, but is alloyed with copper, and sometimes with gold. The copper in the alloy sometimes amounts to 10 per cent. The chief ores of silver, the sulphurets, were noticed in § 53.

91. **Gold.**—This metal has no disposition to unite with oxygen or with other elements, and therefore is found in nature either pure, or alloyed with some other metals, chiefly silver and copper. The only chemical compound formed by gold in nature is with tellurium, termed a tellurid of gold, and this is quite rare. As you have already learned, iron and copper pyrites have often been supposed to be gold by those who are ignorant in regard to minerals. But the distinction is easily made, for gold flattens out on being hammered, and is readily cut, while the pyrites break in pieces on being hammered, and neither of them can be cut in slices, no impression even being made by the knife on the iron pyrites. Besides, the pyrites, on being exposed to a strong heat, emit a sulphureous odor, while the melting gold is odorless. Gold is commonly found in grains or rounded masses of various sizes, but sometimes it is crystallized in the form of the cube or octahedron. Sometimes large lumps have been found. The largest lump that has yet been obtained was from California. It furnished 109 pounds and 4 ounces of pure gold. Gold is diffused in most countries in small quantities; but there are some localities where it abounds, as the Urals, Hungary, Spain, various parts of South America and Africa, but especially Australia and California. In these two last localities it was discovered but recently—in California in 1848, and in Australia in 1851. A locality along the coast of Africa, opposite Madagascar, has been supposed to be the

Ophir of Solomon's time. The Russian mines, that is, those of the Urals and Siberia, were, till of late, the largest sources of gold, but the mines of California and Australia far surpass them; and it is stated by Professor Dana that "the whole product of Europe, Asia, Africa, and South America is far less at the present time than is derived from Australia or the United States."

**92. Modes of Obtaining Gold.**—If the gold be in rock, this is pounded up and sifted, and the sand thus obtained washed in a pan. The gold being seven times heavier than the same bulk of sand or gravel, those portions of the sand which have gold attached to them subside to the bottom of the pan, while the other portions, being lighter, run off with the water. The portions which subside are subjected to the amalgamation process, in which mercury unites with the gold, forming an amalgam, and thus separates the metal from the sand. The gold is then obtained from the amalgam by means which are detailed in § 213, Part II., where the whole process is fully described. This sorting of gold by washing is carried on extensively in nature in the alluvial washings, as they are called. The grains and scales found in the gravel and sand in the beds of streams and on their borders were once in rocks, which were worn away by means that will be made clear to you in another part of this book. The gold, on account of its weight, is always lagging behind the materials that accompany it at the start; and the larger are the pieces of the metal, the less distance are they carried, for small pieces have a larger surface in proportion to their weight for the water to act upon (§ 193, Part I.). For the same reason, a round piece is not carried as far as a scale. When there is a rock in the path of a stream, the pieces of gold collect there, the lighter gravel and sand being carried along by the water. The formation of this "pocket" of gold is very much like the gathering of the golden grains in the bottom of the pan in artificial washing. Sometimes, on the borders of a stream, the material which has been drifted down previously may have much gold in it. To bring it to light, the stream is turned across it, to perform on a large scale the same operation of washing that is done in a small way with the pan. The processes necessary for separating the gold when it is alloyed with silver or copper I will not stop to describe.

93. **Uses of Gold.**—The usefulness of this metal results from its malleability, its ductility, its rich color, its susceptibility of a high polish, and its indisposition to tarnish on exposure. A gold-beater can hammer a grain of gold into a leaf covering a space of 50 square inches, and its thickness is only the 282,000th of an inch. Its ductility equals its malleability, as is shown in the wire of gold lace. It is from these two qualities that gilding is so cheap, and therefore so common, thereby making this costly metal a great convenience to the people at large. For ordinary uses, gold is alloyed with silver and copper to give it the requisite hardness. The standard gold of the United States contains nine parts of pure gold to one of alloy. The word carat, used so often in speaking of the purity of gold in the market, means one twenty-fourth; so that if any specimen is said, for example, to be twenty carats fine, it is meant that the pure gold is to the alloy as 20 to 4, and so of other proportions.

94. **Platinum and its Associate Metals.**—Platinum is seldom found alone, but is commonly mingled with more or less of certain rare metals—iridium, rhodium, palladium, and osmium—and such common metals as iron and copper. It occurs usually in small grains, but sometimes is found in pieces of considerable size. The largest mass that has yet been found weighed 21 pounds Troy. Platinum is often spoken of as being the heaviest substance known, but there is one metal, iridium, that is a very little heavier. Its color is between that of tin and steel, though in the alloys in which it is usually found it is darker than this. It is very malleable,

especially when heated. Its ductility is so great that Dr. Wollaston succeeded in making a wire of it the two thousandths ($\frac{1}{2000}$) of an inch in diameter. In fusibility this metal and mercury stand at opposite ends of the scale. While mercury may be said to melt at 39° below zero, platinum can not be melted in the hottest furnace, as is shown in the crucibles of the chemist that are made of it. The metals iridium, rhodium, and osmium, which are found in company with platinum, are very hard, and are used for points to gold pens. There is a mineral called iridosmine, because composed of iridium and osmium, which is the hardest alloy known. This mineral, which is found in scales, is also used for pointing gold pens.

## CHAPTER VI.

### OXY-SALTS AND HALOID SALTS.

**95. The two Classes of Salts.**—Those salts which are formed by the union of acids with oxyds are called oxy-salts, because they have oxygen in their composition. There is, besides, a class of salts which contain no oxygen, but are composed of a metal and some other element. Common salt is an example. This is composed of chlorine and sodium, and is a chlorid of sodium. This is the principal salt of this class, and hence the name given to the whole class, *haloid*, from two Greek words, *hals*, sea-salt, and *eidos*, like. Some of the oxy-salts have been already treated of in other connections, as the sulphates in the chapter on sulphur and its compounds, from their natural association with the sulphurets. There are also some which will be reserved for another chapter, as the silicates, on the same principle of natural grouping.

**96. Nitrate of Potash.**—This salt, commonly called nitre or saltpetre, is produced in India and South America from the decomposition of animal substances in the soil in hot and dry weather succeeding a rain. It is also produced in the same way in Egypt, and appears in light tufts, which are gathered up by a sort of broom, and the nitre is separated from its impurities by lixiviation, and is crystallized by evaporation. It is found diffused in the earth on the floor of some of our Western caverns, and is apt to appear in crusts or needle-like crystals on their walls. This salt is that one of the ingredients of gunpowder that furnishes the oxygen required for the combustion.

97. **Nitrate of Soda.**—This is very much like nitre; like that, it deflagrates when thrown upon the fire, and it has the same supply of oxygen in it that nitre has; and it would answer the purpose equally well in gunpowder if it did not readily *deliquesce*—that is, gather moisture from the atmosphere. Both this salt and the nitre are used for obtaining nitric acid.

98. **Common Salt.**—This is composed of chlorine and sodium, two elements which never appear in nature uncombined. Its crystals are the cube and its secondaries. A peculiarity in the arrangement of its crystals when they are formed in a solution was shown in § 22. This mineral is very thoroughly diffused in the earth, mostly, however, in solution, nearly one thirtieth of all the water in the sea being common salt. Lakes that have no outlet to the sea are very salt. This is the case with the Great Salt Lake of this country, the Dead Sea, and the Caspian Sea. Over one fifth of the water of the Great Salt Lake is salt, and the proportion is even greater in the case of the Dead Sea. There are famous salt mines where the solid mineral is obtained in Poland, Hungary, Spain, Sicily, and Switzerland. In the extensive mines near Cracow chapels, halls, etc., are excavated far below the surface, their roofs being supported by immense pillars of salt, which, on being lighted up, present a magnificent appearance. In Northern Africa there are hills of salt. In this country vast quantities of salt are obtained by evap-

oration from the water of salt springs. The springs in Onondago county, New York, are the most productive, one seventh part of the water being salt. In hot climates much salt is obtained from sea water by evaporation. The manner in which this is done is described in § 358, Part II.

99. **Borate of Soda.**—Nearly half (47 per cent.) of this salt, commonly called borax, is water. It abounds in a lake in Thibet, where it is literally dug out from the borders and shallow places. As the borax is deposited from the water, the holes made in gathering it are soon filled up. This lake is at so high an elevation that it is frozen over the greater part of the year. The acid which enters into the composition of borax, boracic acid, is sometimes found in the vicinity of volcanoes, exhaling from springs in the earth. It is present in the hot vapors of the lagoons in Tuscany, from which it is obtained by letting the vapor pass into cold water to condense it, and then allowing the solution to evaporate. This leaves the boracic acid in large crystalline flakes.

100. **Carbonate of Lime.**—This mineral, in the forms of common limestone, chalk, and marble, constitutes about one seventh part of the crust of the globe, and therefore will be brought quite largely to your notice in the geological part of this book. There are mountains made of it, and there are extensive deposits of it in layers of rock. It appears also in smaller quantities in many other forms, some of them exceedingly beautiful. It is readily distinguished from other minerals by dropping some acid, as the sulphuric, upon it, this occasioning an effervescence by uniting with the lime and setting free the carbonic acid, as already stated in § 60. Exposed to a red heat the carbonic acid is driven off, and we have the lime left alone. This is what is done in the lime-kiln, and it is in this way that the lime used for making mortar and other purposes is obtained. *Hydraulic lime*, which has this name because it will "set" under water, is made from

limestone in which there are some silica and clay, and sometimes magnesia. *Marl*, which is of so great use in agriculture, is a mixture of carbonate of lime with clay. When carbonate of lime is deposited from the waters of a mineral spring it is called *calcareous tufa*. Crystallization is prevented in this case from the constant motion of the water. Chalk, though abundant in England, France, and many other countries, is not found at all in this country, though the other common forms of carbonate of lime are present in large quantities.

101. **Calcareous Spar.**—This name is usually given to the crystallized varieties of carbonate of lime. The crystals vary exceedingly from the variety of their secondary forms. Some fanciful names are given to some of them from their peculiar shapes, as dog-tooth spar and nail-head spar. They are various in color, commonly white or light gray, or reddish or yellowish, and sometimes wine-yellow, red, rose, or violet. The crystals of *Iceland spar*, so called because they were first brought from Iceland, are transparent, and they exhibit double refraction—that is, objects seen through the crystal appear double. *Satin spar* has its name from the satin lustre which its beautifully fibrous arrangement presents. There is a form of carbonate of lime, called aragonite, that crystallizes after a different plan from that of calcareous spar, and it is much harder.

102. **Marble.**—This is a granular limestone, the grains being imperfect crystals from their encroaching on each other during their formation. The finest varieties are called statuary marble, the best coming from Carrara, in Italy, the island of Paros, and some other localities in the same quarter. The coarse kinds are the common marbles, which are either white or clouded with various coloring substances. In some marbles there are shells or corals, these being composed of the same mineral with the marble itself. Indeed, it is supposed by some that all marble and other forms of carbonate of lime are made

of the shells and skeletons of animals, and that it is by the agency of fire chiefly that their forms have been to so great an extent destroyed, the particles that composed them having taken on a crystalline arrangement in the marble. There is one mode of arrangement, differing from the ordinary one, which I will notice in passing. We see it in both marble and common limestone. It gives to marble a grayish color, and an appearance resembling the roe of a fish, from the rounded dots with which it is speckled. It is called *oolitic* marble, and the limestone which is composed of these round grains is called *oolite*. This term comes from the Greek word *öon*, egg.

103. **Stalactites and Stalagmites.**—In many limestone caves the carbonate of lime accumulates overhead in shapes like icicles, and these are called stalactites. Accumulations also occur on the floor of the cave similar in form to those which we sometimes see on the ground under dripping icicles, and these are called stalagmites. The resemblance in form in the two cases is very striking. For the chemical explanation of these formations I refer you to § 307, Part II. Sometimes the stalactites and stalagmites meet, forming pillars, and these may, in the course of time, become very large, so that, where the roof is high, the appearance by torch-light is magnificent. In Weyer's Cave, which is a mile and a half in extent, and in some parts forty feet high, many of the stalactites being of a delicate white color, in contrast with the blue limestone of the walls, the scene is exceedingly grand and beautiful.

104. **Magnesian Carbonate of Lime.**—This mineral, called dolomite, is composed of carbonate of lime and carbonate of magnesia. Its crystals, which are rhombohedrons, are often curved, as seen in Fig. 3. Much of the coarser kind of white marble in use for building is dolomite in the granular or imperfectly crystallized form.

**105. Fluor Spar.**—This is a fluorid of calcium, composed of fluorine (one of the chlorine family) and calcium, which is the metallic base of lime. Neither of these elements is ever found in nature. The crystals of fluor spar are the cube and the octahedron, and their secondaries. Their colors, which are bright, are white, light green, purple, and yellow; sometimes rose-red and sky-blue. Fluor spar phosphoresces brightly when put upon hot iron, giving out various colors. It is an abundant mineral in Derbyshire, England, and it is therefore often called **Derbyshire spar**. In the massive* or granulated form it is susceptible of a high polish, and vases, candlesticks, and various articles of ornament are made from it. This mineral is present in very small quantities in the teeth and bones of animals; and as there must be a supply of it for these structures from some source, it is found in some plants, which, of course, have imbibed it, as is the case with silex, from the earth.

**106. Apatite.**—Most of this mineral, whose beautiful crystals present much variety of form, is a compound of phosphoric acid and lime—that is, a phosphate of lime. But there is intimately incorporated with this fluor spar, and a very small quantity of the chlorid of calcium.

**107. Salts of Magnesia.**—The *sulphate of magnesia* was noticed in § 61. The *carbonate* appears like some of the varieties of carbonate of lime and dolomite. Sulphate of magnesia (Epsom salt) is often made from it. This is done by means of sulphuric acid, the acid seizing the magnesia and uniting with it, the carbonic acid gas, of course, flying off as it is released. The *hydrate of magnesia* is a compound of magnesia and water, the latter being nearly one third of the whole. *Borate of mag-*

---

* The term *massive* is applied to minerals that are imperfectly crystallized—that is, having parts of crystals huddled together in a confused mass. The crystals thus massed together may be granular, or fibrous, or laminated. We have a familiar example of the granular in marble, and of the laminated in slate rocks.

*nesia*, called *boracite*, is a compound of boracic acid and magnesia.

108. **Chlorid of Ammonium.**—This was formerly supposed to be composed of muriatic acid and ammonia, but modern chemistry has discovered that it is a compound of chlorine and ammonium, a metal which has never yet been seen, but whose existence has been satisfactorily proved, as shown in § 230, Part II. The common name of this mineral is sal ammoniac. It occurs in the vicinity of volcanoes, and is the result of volcanic action.

109. **Alum.**—Common alum is a double salt, a combination of sulphate of alumina and sulphate of potash. But there are other alums—a *soda* alum, in which there is sulphate of soda in place of the sulphate of potash; a *magnesia* alum, in which there is sulphate of magnesia; an *iron* alum, in which there is sulphate of iron; an *ammonia* alum, in which there is sulphate of ammonia; and a *manganese* alum, in which there is sulphate of manganese. Then there is what is called *feather alum*, which is not a double salt, but a simple hydrous sulphate of alumina, and therefore can not, strictly speaking, be called an alum. This is more abundant in nature than any of the true alums. This salt, the potash alum and the iron alum, often impregnate the rocks called clay slates, and when thus charged they are termed aluminous slates or shales. Alum is often obtained from these rocks by lixiviation.

110. **Phosphates of Alumina.**—Wavellite, a mineral which is found adhering to rocks in small hemispheres, is a phosphate of alumina, having combined with it a small amount of fluorid of aluminum. Turquois, an opaque greenish-blue mineral, much used for ornament, is a phosphate of alumina, having combined with it phosphates of copper and iron in small quantities. This gem is often imitated, and the counterfeits are sometimes so good that they can be detected only by chemical tests.

## CHAPTER VII.

### EARTHY MINERALS.

111. **Composition.**—A large proportion of the earthy minerals are silicates of the earths alumina, lime, magnesia, etc. These earths, as you learned in Part II., are oxyds of metals, and, united with silica (that is, silicic acid), they form salts called silicates. The different kinds of glass are artificial silicates, many of them very beautiful, but surpassed greatly in beauty by many of the natural silicates found in the rocks. Many of these silicates are very compound, the silica being united with several bases at the same time. Some examples of double salts (salts in which one acid is united with two bases) were brought to your notice in the alums, and some other salts, in the preceding chapter. But many of the earthy minerals are much more compound than this. Some have not only several bases, but more than one acid. Thus, in that beautiful azure-blue mineral, lapis lazuli, we have both silicic acid and sulphuric acid. Some of the earthy minerals, on the other hand, have a very simple composition. Thus the splendid sapphire and common emery are both the pure earth alumina, that is, oxyd of aluminum; and then we have, as you will see, many varieties of quartz, which is pure silica, or silicic acid.

112. **Silica.**—This substance constitutes about 45 per cent. of the crust of the earth, some rocks being entirely composed of it, but more of them having it in combination or in mixture. In the granite we have it in both these states, for the quartz is pure silica; and while this is mixed with two other minerals, mica and feldspar, in these minerals, as you will see in another part of this

chapter, the silica is in chemical combination with other substances, forming silicates. Silica is a very hard substance, scratching glass readily, but inferior in hardness to the diamond. It is insoluble, and can not be melted, even in the strongest heat that can be obtained by the blow-pipe. It appears in greater variety of form and color than any other mineral, but its qualities are so marked that it is easily recognized. As it is so abundant in the rocks, and yet hard and insoluble, its fragments abound, most of the pebbles in gravel and in the common soil, most of the sand, and much of the hard grains even in what we call fertile earth, being silica. Of course, in every case where there is a decided color there is something besides the silica—that is, there is some coloring substance. But where the mineral, when colored, is clear and transparent, there is so little of a foreign substance, and it is so intimately mixed or combined with the silica, that it can hardly be regarded as an impurity. But sometimes, especially when the mineral is opaque, the impurity is palpable—oxyd of iron, clay, etc. Flint is one of those opaque forms. This, which was formerly in such wide use in muskets and in the common tinder-boxes, but is now superseded by percussion caps and lucifer matches, is quite an abundant mineral, and is extensively used in pottery.

113. **Quartz.**—There are said to be three varieties of quartz—the *vitreous*, the only one which appears in crystalline forms, its name coming from the fact that its fracture is glassy; the *chalcedonic*, generally translucent, with a waxy lustre, often exhibiting several colors, generally fantastically arranged; and the *jaspery*, of a dull red, sometimes yellow color. But in common language, even among mineralogists, it is only specimens of the first variety that are usually spoken of as quartz; and, on the other hand, the two terms silica and quartz are often used as being synonymous. The crystals of quartz are usually six-sided prisms, terminated with six-sided

pyramids, but modified so as to present much variety. Some of the forms are represented in Fig. 22. Some of the crystals have the pyramidal terminations at both ends.

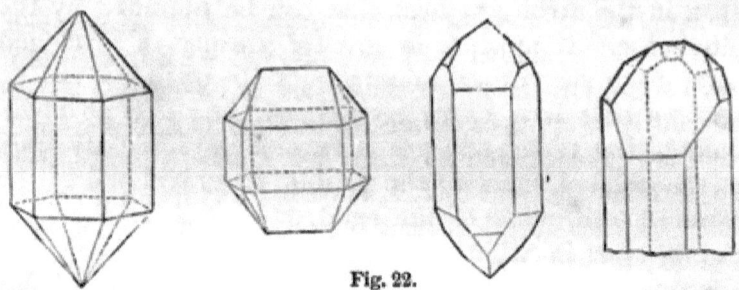

Fig. 22.

Nothing can be more pure and transparent than clear, uncolored, limpid quartz. The common name for it is rock crystal. It is said that it was to this mineral that the ancients first gave the appellation of crystal, from its resemblance to perfectly clear ice, the Greek word for ice being *krustallos*. It is often used in jewelry, and also to make lenses for optical instruments. The *amethyst*, so called, is a variety having a purple or red color from the presence of some oxyd or oxyds. The rich purple specimens, so much prized for jewelry, derive their color from the oxyd of manganese. Sometimes the crystals of quartz have a light yellow color, and then are called false topaz. *Smoky* quartz appears with various degrees of color. Crystals with the lighter shades are often very beautiful, and are sometimes used in jewelry. Sometimes quartz is filled with golden-yellow spangles of mica; but the artificial imitations of this mineral, contrary to the general fact that nature excels art, transcend in beauty the original. The colors of what is called *ferruginous* quartz, yellow, brownish-yellow, and red, are produced from oxyd of iron, and hence its name, *ferrum* being the Latin word for iron.

114. **Chalcedony.**—This kind of quartz, as already stated, does not appear in crystalline form, but in spher-

ical and nodular masses. In its various forms it is used for making various articles—cameos, snuff-boxes, buttons, marbles, etc. The carnelian, which is so familiar to you, is of a bright red color, sometimes yellow. There is one variety of an apple-green hue, the color being given to it by nickel, the metal found alloyed with iron in meteorites (§ 71). In *agate* the colors are arranged either irregularly in spots or clouds, or in regular layers around a centre. Sometimes the lines are zigzag, like the lines of a fortification, and then it is called fortification agate. Sometimes the oxyd of iron is arranged in this mineral in moss-like branches, giving it the name of *moss agate*. In the *onyx* the differently colored material is in horizontal layers, the colors being commonly a light brown and a white. This is the mineral which is used for cameos, the figures being cut in one layer, the other layer furnishing the background. The variety of chalcedony called *cat's eye*, of a greenish-gray color, has internal reflections of light which resemble those of the eye of a cat.

115. **Opal.**—This mineral belongs to neither of the varieties of quartz, differing from them in composition by containing water, the water varying in different specimens from 5 to 12 per cent. The colors of opal vary much from the presence of coloring matters, which, especially in the dark and thoroughly opaque specimens, may be regarded as impurities. The color of the *noble opal*, so highly prized as a gem, is commonly milky, and has a play of brilliant but delicate internal reflections. In the *fire opal* these reflections are a *fire-red* in color. Opal has not as much hardness as quartz.

116. **Silica in Solution.**—I have stated that silica is insoluble. It is ordinarily so. But sometimes it is dissolved in water by means of the potash in it, as explained in § 265, Part II. Thus, in the waters of the geysers and some other hot springs, there is silica in considerable quantity. Here, undoubtedly, heat aids the solution, but it is not its principal cause, for silica is dissolved in water

at ordinary temperatures to some extent in all parts of the globe, as is shown by its deposition in grasses and other plants, and in the textures of animals also, especially some of those minute animals called infusoria, as you will see in another part of this book. These deposits are of course made from solution, the solvent being the sap in vegetables and the blood in animals.

117. **Silica in two States.**—There are two states, then, in which silica exists, the soluble and the insoluble. Most of it in the world is in the insoluble state, in the rocks and pebbles, and sand and grains in the soil; but some of it is changing continually from this state into the soluble, and as continually passing back again by deposition. Thus, by the agency which I have mentioned, some of the silica in the soil (which came originally from the rocks) is dissolved in the water, and finds its way into the plant by the sap, where it is deposited in the texture, and when deposited it is insoluble again. If the plant decay or is burned the silica is returned again to the earth. The same succession of changes we have also in relation to some animals. And it is supposed that the insoluble silica which we have in chalcedony, jasper, etc., is a deposit from the soluble state.

118. **Petrifactions with Silica.**—When wood decays in water which contains considerable silica in solution, petrifaction occurs; that is, as the particles of wood are removed in the process of decay, particles of silica are deposited in their place. Of course the silica is deposited from a solution of it that penetrates the wood. The arrangement of the wood is preserved, so that it appears as if the wood had been turned into stone. Some specimens are exceedingly beautiful when sawn across what was the grain of the wood and polished.

119. **Silicates of Lime.**—These are of so little importance that I need not describe them. The same is true of the boro-silicates—that is, compounds of silicic and boracic acids with lime combined together in one mineral.

120. **Silicates of Magnesia.**—These are of two classes, the hydrous and the anhydrous. In them all, which are numerous, there is not merely a silicate of magnesia, but there are also other oxyds, some in one and some in another, as oxyd of iron, alumina (oxyd of aluminum), lime (oxyd of calcium), the oxyd of manganese, etc. These additional oxyds are generally small in amount, but in some cases they make quite a considerable portion of the mineral. I will notice only the most important of these silicates.

121. **Talc.**—This is one of the softest of minerals, being easily cut with a knife. It is commonly of a light green color, and has an unctuous feel. One of its varieties, scaly talc, is the French chalk so familiar to us. Another, the soapstone (steatite), is found in extensive beds, and is used for many purposes—fireplaces, linings for stoves, etc. It is, you know, rather soft, but heat hardens it. Powdered soapstone is used for lessening the friction of machinery.

122. **Serpentine.**—This, when so pure as to be called precious serpentine, is almost a pure hydrous silicate of magnesia, the only additional oxyd in it being that of iron in the small quantity of 0.2 per cent.—that is, $\frac{1}{500}$th of the whole. When polished it is a very rich-looking stone, the color being green. Other varieties contain more of the oxyd of iron. There are some rocks composed wholly of serpentine, and others that have serpentine mingled with other minerals. The *verd-antique* marble is granular limestone with serpentine scattered through it. Serpentine was so named from its resemblance to the skin of a serpent, being streaked or spotted.

123. **Chlorite.**—This is a dark green hydrous silicate of magnesia, alumina, and iron, there being in it 17 per cent. of alumina and 34 of magnesia. In some parts of the earth there are extensive deposits of this mineral, and also of a slaty rock, which is called chlorite slate, because chlorite is its chief constituent.

**124. Pyroxene.**—This is a very common mineral, and has many varieties. The colors also are various, being shades of green from the lightest to the darkest. One variety is even white. One of the additional oxyds is lime, which differs in quantity very much in the different varieties. Pyroxene occurs in various rocks—granite, limestone, the lavas, etc.

**125. Hornblende.**—This constitutes a large part of some rocks, as trap and some kinds of slates, giving to them great toughness. Like pyroxene, it contains lime, and some of its varieties it is difficult to distinguish from that mineral. Some of its varieties are beautiful. The famous asbestos is a very remarkable one, being arranged in such slender silky fibres that it may be woven like cotton or linen into cloth. This cloth is incombustible, and, when soiled, can be effectually cleaned at once by putting it into the fire. The Greenlanders use asbestos for lamp-wicks, and in ancient times it was used for this purpose in the temples, its incombustibility being thought to give it a sacred character. Amianthus is a variety of asbestos which has a satin lustre. In Siberia and Spain, gloves, ribbons, and purses are made of it. The difference between the tough hornblende that we find in rocks and this very delicate variety of it is no greater than that between common wood and the delicate varieties of wood which we see in the forms of cotton and linen, and in the textures of leaves, flowers, etc., as noticed in § 511, Part II.

**126. Alumina.**—This is an oxyd of aluminum, a metal which is never found in nature, but is obtained artificially from its compounds. Though this metal has been little known till lately, it is coming into considerable use, especially among the French, on account of qualities which are noticed in § 238, Part II. The oxyd alumina is familiar to us in the form of *emery*. This is granular. But it also appears in various crystalline forms, and, when clear, is exceeded in costliness only by the diamond. It is the *sapphire*, and is of various colors. It is only the blue crystals that commonly receive this name, while a red crystal of it is called the Oriental ruby, and a yellow one the Oriental topaz. This mineral is in hardness inferior only to the diamond, scratching quartz quite easily. Alumina is the essential ingredient of all clay, and is therefore very abundant, entering into the composition of many rocks, and forming a constituent more or less of the soil. It is calculated that aluminum, its base, is one twentieth part of the crust of the earth.

127. **Spinel.**—This mineral is composed essentially of alumina and magnesia intimately combined, there being small quantities in it of other substances, as oxyd of iron, silica, and chromic acid. The finely-colored crystals are prized as gems in jewelry, and the red is the *common ruby*, in distinction from the Oriental, which is a sapphire. There are some varieties of spinel in which the oxyd of zinc is a prominent ingredient.

128. **Silicates of Alumina.**—There are a few silicates of alumina that have no other substance in combination, one of which is named Sillimanite, in honor of Professor Silliman, sen., of Yale College. But most of the silicates of this earth are quite compound, having various oxyds and other substances combined with them, as lime, magnesia, potash, soda, oxyd of iron, lithia, oxyd of manganese, etc. I will notice a few of those which are most prominent in interest.

129. **Feldspar.**—This mineral, which is a silicate of alumina and potash, is one of the three crystalline constituents of granite, and its partially formed crystals are easily distinguished in the coarser specimens of that rock. It is not as hard as quartz, and is more brittle. Its crystals are commonly four or six-sided prisms, and have a pearly lustre. Two of its usual crystals are represented in Fig. 23 (p. 64). The colors of this mineral are commonly white, milk-white, gray, and flesh-colored, but sometimes the crystals are violet and green. Some-

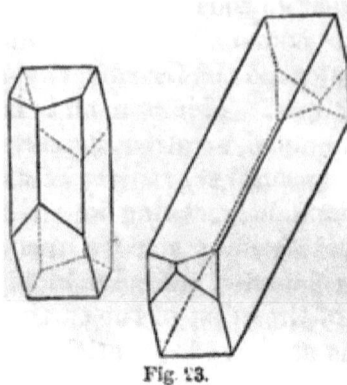

Fig. 23.

times feldspar is opalescent—that is, resembling opal in lustre, and some specimens of this variety are used in jewelry. Sometimes it is iridescent, from the glistening of minute crystals of specular iron contained in it. Feldspar is used largely in the manufacture of porcelain, the clay which comes from the decomposition of the feldspar being called *kaolin*. *Albite*, so called from the fact that white is its most common color, differs from feldspar chiefly in containing soda instead of potash. It takes the place of feldspar sometimes in the rocks. It does so in granite, and albite granite differs from feldspar granite in having a lighter color, owing to the whiteness of the albite. *Labradorite* is another mineral sometimes found in granite. It differs from feldspar and albite in having a large percentage of lime.

130. **Mica.**—This is another of the three constituents of granite. It has chiefly the same substances in its composition as feldspar, but there are added small amounts of oxyd of iron, fluoric acid, and water. It has many varieties, but that which is so familiar to us under the common but inappropriate name of isinglass consists of extremely thin plates, transparent, with a bright lustre, the colors being white, yellow, gray, brown, and blackish green. It is very elastic, being unlike talc in this respect, though resembling it in some others. It is used for doors in lanterns and windows in ships, where glass, from its brittleness, is liable to be broken. As it bears heat well it is used much in stoves. In Siberia, where it is obtained in large quantities, it is used in place of glass. Plates of it three feet square have been brought from that country, and they have been obtained of the

same size in our own country in several localities in New Hampshire.

131. **Garnet.**—In this mineral there are combined the silicates of alumina, lime, iron, and manganese, and the various colors of the different specimens generally come from the difference in proportions of these ingredients. The most common color is deep red. In an emerald-green variety the color is caused by oxyd of the metal chromium. Clear garnets of a deep red color are highly prized in jewelry.

132. **Tourmaline.**—The usual form of the crystals of this mineral is a prism terminating in a low pyramid. They are commonly long, and the sides are apt to be furrowed. The most common colors are black, blue-black, and dark brown; but, besides these, there are wide varieties, as bright and pale red, grass-green, yellow, etc., even to white. Though alumina is the chief base in all the varieties, the differences in the composition of some of the varieties are considerable. For example, in a black variety there was found to be 24 per cent. of oxyd of iron, while in a red variety there was no iron at all, but 5 per cent. of the peroxyd of manganese, a frequent cause of this color in minerals. Some red and yellow tourmalins are of great value as gems. A Siberian tourmaline of the red variety, now in the British Museum, is considered to be worth £500.

133. **Topaz.**—This is said to be a silicate combined with a fluorid. If so, it must be a combination of silicate of alumina with fluorid of the metal aluminum, for there can be no such thing as a fluorid of alumina any more than there can be a chlorid of soda. In explanation of this I refer you to § 352, Part II. The clear crystals of topaz are used in jewelry, and it is a curious fact that the color can be altered for this purpose by exposure to heat. The Brazilian topaz can be thus made so to resemble the real rose-red ruby that it can not be distinguished from it except by an electrical test. When the

most clear, pellucid pebbles of topaz are cut with facets and set, they appear by daylight precisely like diamonds.

134. **Lapis Lazuli.**—In this beautiful azure-blue mineral we have a compound of silica, alumina, soda, lime, iron, sulphuric acid, sulphur, chlorine, and water. It is plain that the silicic acid and the sulphuric acid must form with some of these oxyds a silicate and a sulphate, which are intimately combined together, but how some of these substances are combined we know not. The color is attributed to sulphuret of sodium. There are some specimens of this mineral whose composition differs somewhat from that given above.

135. **Beryls and Emeralds.**—A beryl is a silicate of alumina and glucina, while an emerald is the same, colored with oxyd of chromium to the amount of less than one per cent. The color is mostly green—pale in the beryl, but decided and rich in the emerald. The best emeralds are brought from Granada, where they are found in dolomite (§ 104). One, weighing six ounces, cost the owner, Mr. Hope, of London, £500. Emeralds of larger size, but of less beauty, are obtained in Siberia. One in the royal collection of Russia measures four and a half inches by twelve. The finest beryls come from Siberia, Hindostan, and Brazil. Some beryls of very great size have been obtained in this country, but they are seldom transparent. A hexagonal prism was found in Grafton, New Hampshire, of such enormous size that it weighed 2900 pounds. It is four feet in length. The *chrysoberyl* is a combination of alumina and glucina, sometimes with a little iron. The color is indicated by the name, which comes from two Greek words, *chrysos*, yellow, and *beryllos*, beryl.

136. **Zircon.**—This mineral, which is a silicate of the earth zirconia, has various colors. The clear crystals are in common use in jewelry. When they are red they are called hyacinths. This variety has some resemblance to the red spinel. Zircon is used in jeweling watches.

## CHAPTER VIII.

### ROCKS.

**137. Connection of Mineralogy with Geology.**—In this chapter we shall pass over ground that connects mineralogy with geology. We have looked at minerals in detail and in their separate condition, merely alluding to them occasionally as being masses, but now we are definitely to consider these masses preparatory to a view of them as a whole, making up the crust of the earth. It is such a view that constitutes Geology, the term being derived from two Greek words, *ge*, earth, and *logos*, discourse.

**138. Definition of Rock.**—The word rock, in common language, is applied only to such mineral aggregates of considerable size as are solid; but the geologist gives it a wider meaning. He speaks of all such aggregates as rocks, whether they are solid or made up of loose material. Thus the clay slate, and the strata of clay mud, which, by long-continued pressure, or this and heat together, might ultimately become clay slate, are both equally regarded as rock. The same may be said of sandstones and deposits of sand. The geologist, however, often uses the word rock in its ordinary sense, the context showing in which sense he does use it.

**139. Elementary Substances in the Rocks.**—There are between sixty and seventy elements, but only nine of them enter to any great extent into the composition of the rocks, viz., oxygen, silicon, aluminum, calcium, magnesium, potassium, sodium, iron, and carbon. Of these only one, oxygen, is a gas, but that is so abundant that it is supposed to constitute nearly or quite one half of all the ponderable matter in the globe. These elements, it is stated by Dana, make up $\frac{977}{1000}$ths of the rocks as a whole. But there are some other elements which enter into the composition of some rocks. Sulphur exists in sulphate of lime or gypsum, which in some parts of the earth occurs as a rock in beds of considerable extent. This is the only compound containing sulphur of which any rocks are made. Sulphurets, as you saw in Chapter IV., are abundant, but they are only ores in rocks, and do not enter into their composition. Hydrogen is another element that is present in some rocks. It is present in the water, which is, as you have seen, a component of hydrous minerals, and some of these, as gypsum and serpentine, form rocks. The amount of water in gypsum is 21 per cent.

140. **Compounds in the Rocks.**—The elements of which I have spoken do not form rocks as elements, but as compounds. Thus oxygen forms with silicon a compound, silica or silicic acid; and this silica constitutes rock of itself (quartz rock); or makes a part of a mixture of rock, as in the quartz of granite; or, uniting with some oxyds, makes silicates, as in the feldspar and mica of granite, in serpentine, in chlorite, etc. The most important of the compounds which compose the rocks of the globe are silica, carbonate of lime, and various silicates, viz., feldspar, mica, hornblende, and pyroxene. Besides these, the only minerals that have any large share in making up the earth's crust are chlorite, serpentine, talc, gypsum, and coal. All these minerals have been sufficiently described in previous chapters.

141. **Crystalline and Uncrystalline Rocks.**—It is plain that marble is crystalline, although none of the individual crystals, as they were crowded together in their formation, was completed. The glistening is occasioned by the small portions of the faces of the crystals which reflect the light. The same incomplete crystalline structure is seen in the three minerals that compose granite, as we look at a fractured face of that rock. But sand-

stones were evidently formed by a different process. Here we have grains *mechanically* mixed, as in a layer of sand on a sea-shore, and then in some way becoming solid rock. The grains themselves may have been formed at the outset when they were a part of some other rock by a crystallizing process; but after the original rock became broken and worn in one way and another into these grains, they became arranged by the agency of water in layers, and changed into solid rock by means which we shall hereafter consider.

142. **Rocks Composed of a Single Mineral.**—Rocks sometimes are composed of one mineral alone. Pure limestone, in its three varieties of common limestone, chalk, and marble, is of this character, being simple carbonate of lime. Sometimes, though a rock may be a single mineral throughout, that mineral may not be simple in its composition. Thus the rock called dolomite is a magnesian carbonate of lime, or, in other words, carbonate of lime and carbonate of magnesia intimately combined, so as to make one mineral. Other examples of rocks composed of a single mineral are quartz rock, gypsum (sulphate of lime), serpentine, etc.

143. **Rocks Composed of more than one Mineral.**— Most rocks are constituted by a mixture of different minerals. Granite is a familiar example, being a mixture of quartz, feldspar, and mica. Sometimes the ingredients are so finely mingled that, different from granite, the composite minerals can not be at all distinguished. Pudding-stones are aggregates or conglomerates of pebbles, united by some cementing mineral, most commonly silica, or oxyd of iron, or carbonate of lime. The pebbles may be of one kind or of different kinds, and they vary much in size. When the conglomerate contains angular pieces instead of rounded pebbles, it is called *breccia*. Usually the pebbles or pieces are granite, or quartz, or carbonate of lime, and the conglomerate is accordingly said to be respectively granitic, quartzose, or

calcareous, this latter term being commonly applied to all compositions in which there is limestone.

**144. Stratified and Unstratified Rocks.**—It is manifest to the most superficial observer that some rocks are in layers or strata, while others have a very different arrangement. A full consideration of the modes of construction of the different forms of rock belongs to another part of this book. Suffice it to say now that the materials of the stratified rocks were assorted and laid down by water, and then by some means became solidified, while the unstratified were made under the influence of strong heat.

**145. Silicious, Argillaceous, and Calcareous Rocks.**—Stratified rocks are divided into three classes, according to their composition. The first class is the silicious or arenaceous. This is the sandy division, called arenaceous from *arena*, the Latin word for sand, and silicious, because the grains of which the rock is composed are silica, or silex, as it is commonly called. The second class, the *argillaceous*, are made of clay, which is a mixture of silex and alumina, commonly in the proportion of about three of the former to one of the latter. The name comes from *argil*, a term applied technically to alumina, and probably to clay. It is a characteristic of rocks of this class that they give out a peculiar earthy odor when breathed upon, which is not owing to the presence of alumina simply, but to a combination of this with some oxyd of iron. *Calcareous* rocks are those which are composed of lime and carbonic acid. These three classes of rocks are seldom found pure, but they run into each other. For example, there are sandstones which are not wholly silicious, the silicious grains being united together by carbonate of lime. Whether any silicious or argillaceous rock has the carbonate of lime in it can at once be ascertained by the application of sulphuric acid (§ 60.)

I pass now to consider some of the individual rocks.

146. **Granite.**—Ordinary granite is a mixture of quartz or silex, mica, and feldspar. There is great variety in this rock, according to the varying proportions of these minerals, and each variety is termed feldspathic, micaceous, or quartzose, as one or the other mineral predominates. There are other variations also. Sometimes hornblende is present in place of mica, and then the rock is called *syenite*. This is even more durable than common granite. Its name comes from Syene, in Upper Egypt, the locality from which most of the stones of the ancient Egyptian monuments were obtained. These are not, however, true syenite, but a red granite, containing considerable dark-colored mica. The rock of Mount Sinai is real syenite. When albite is in the place of feldspar, the rock is called albite granite (§ 129). When talc replaces it, it is called *protogine*. When the feldspar appears in the granite in large crystals, it is called *porphyritic* granite. There is a peculiarity in the crystalline arrangement of one kind of granite which produces figures over its surface like small Oriental characters, and hence this variety is styled *graphic* granite. In Fig. 25 is given the surface of a slab of this granite, and in Fig. 24 its end. The colors of granite vary much, but it

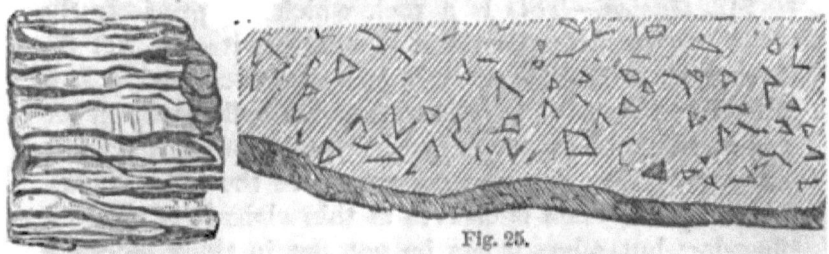

Fig. 24. Fig. 25.

is usually grayish, white, or flesh-colored. Rocks of the granite family are widely disseminated in the earth, often forming extensive mountain ranges. In some parts of the Andes granite rises to the height of 12,000 feet. In the Alps, the Aiguille de Dree is a solid spire of granite 4000 feet high. There is much granite of different kinds in New England, Massachusetts being truly the granite state of the Union, though New Hampshire is generally called so.

147. **Uses of Granite.**—Granite, on account of its durability, is one of the most valuable materials for buildings and monuments. It was much used by the ancients, and there are obelisks of granite in Egypt that have been exposed to the weather for three thousand years, and yet are in good condition. Commonly, the finer is the texture of granite, the more durable is it. Syenite, which is so durable, is of a fine texture, but it owes its durability partly to the toughness which hornblende imparts (§ 125). If there be pyrites or any other ore of iron disseminated through the rock, it impairs its value essentially, because by decomposition rust is produced, defacing and often crumbling the stone. Some granites, which present a good appearance on being taken out of the quarry, very soon crumble on continued exposure to the air without any obvious cause. This shows the necessity of proper examination in selecting granite for building. As feldspar contains a large proportion of alumina, feldspathic granite is much used in obtaining the kaolin for making porcelain.

148. **Gneiss.**—This is a rock which has precisely the same mixture of minerals in it that granite has, but the arrangement is different. It is a stratified rock. It is foliated; that is, it is in leaves, and is split easily in different thicknesses, varying from a few inches to a foot or more. This cleavage occurs where the mica is most abundant. When it cleaves in thin slabs it is used for flagging, but when it can be got out in thick blocks it is used for building. Some of the buildings of Amherst College are constructed of this material, and present a fine appearance. There are many quarries of it in Massachusetts, and some in Connecticut.

149. **Slates, Shales, and Schists.**—These terms are often used as being synonymous; but when they are used

as having different meanings, the distinctions are these: *Slate* is the term used when the rock splits into thin laminæ, and when the word is used alone it means clay slate. But there are mica slates, hornblende slates, etc. The word *shale* is used when the rock, though slaty, is more brittle than slate, and so crumbles readily. *Schist* includes rocks so coarse that it would not be proper to call them slates or shales. It comes from the Greek word *schistos*, divided or split.

*Mica* slate, or schist (for it is called by both names), has the same ingredients with gneiss, but has less of feldspar and more of mica in it. The scales of mica give a glistening appearance to the surface of the slabs, which are used for flag-stones, hearth-stones, and door-steps. Furnaces are sometimes lined with them. Scythe-stones are made out of varieties that have a fine grain.

*Hornblende* slate, or schist, is more durable than mica slate, from the toughness which the hornblende gives it (§ 125), and is therefore very valuable for flagging.

*Talcose* slate, or schist, is brittle, but is used for firestones.

*Clay* slate has about the same composition with mica slate, but the ingredients are so finely mixed that they can not be distinguished from each other. The colors are commonly bluish, greenish, gray, or reddish. Slate is used for roofing, and for making drawing-slates, pencils, etc.

150. **Quartz Rock.**—This is composed of quartz, either in the granular or arenaceous (sandy) form. There are varieties which result from the admixture of other substances, as mica, feldspar, etc. In these varieties there is regular stratification, especially in the micaceous, which often cleaves into slabs, like gneiss and mica slate; but when the rock is unmixed granular quartz there is no obvious stratification. Quartz rock is used for flag-stones, hearth-stones, and fire-stones, and in the form of cobblestones for paving. The fine sand into which this rock

sometimes crumbles is employed in making glass, and in sawing and polishing marble. The sand of our sand-paper comes from this source. The *buhrstone*, which is made into mill-stones, is quartz rock that is full of little spaces or cells, these making the surface rough with sharp minute edges crossing each other in every direction.

151. **Sandstones.**—These usually consist mostly of silicious sand, some of the varieties of quartz rock approaching very nearly to them in character. Sometimes there is much clay in their composition, and then the rock is called an argillaceous sandstone. Sometimes there are silicious pebbles imbedded in the rock, making it in part a pudding-stone, and in that case, if the rock be very hard, it is called a grit rock or mill-stone grit. Some sandstones that readily split into comparatively thin layers are much used for flagging-stones. Sandstones have dull colors of various kinds, from white, through grades of yellow, red, and brown, even to black, these being caused, of course, by various substances combined with the silex or quartz. When sandstone has a fine, even grain, it is a very beautiful building material. More caution is necessary in selecting sandstone for building than granite, because varying circumstances of exposure to air, moisture, and heat have so much influence upon it. The sandstone obtained chiefly from New Jersey and various localities in the valley of the Connecticut, now so much used in almost all parts of the country to which it can be carried by water, is generally an excellent material. That used in building the Capitol at Washington, which came from the Potomac, is, unfortunately, an inferior article.

152. **Trappean Rocks.**—This appellation is given to certain rocks which have such resemblance to each other as makes it proper to group them together. The term trap comes from a Swedish word meaning stair, the rocks of this class often presenting to the eye an appearance like steps. The term *greenstone* is applied to the com-

pact rocks of this class in which hornblende is largely present, and imparts a green color. *Basalt* is much like greenstone, but is black or grayish-black. It contains small grains of a silicate of magnesia and iron, called *olivine*, from its olive-green color. *Trachyte* is a grayish-white rock composed of feldspar, hornblende, and mica. *Clinkstone* is a grayish-blue feldspathic rock, which bears this name because it rings like iron when struck with a hammer. The term *porphyry* is used in reference to the structure of rocks rather than their composition. It is applied to any feldspathic rock that has crystals disseminated through it. It is called greenstone porphyry, basaltic porphyry, etc., according to the material which makes up the body of the rock, or, as it is commonly expressed, the *base* of the rock. What the ancients called porphyry is a rock having a base of compact feldspar, with imbedded crystals of feldspar of various sizes, from a very small size up to a length of three fourths of an inch. The name *amygdaloid* refers also to structure. It comes from two Greek words, *amygdale*, almond, and *eidos*, like, and is applied to any rock of the trap family that has in it rounded cavities filled with some mineral different from the base of the rock. The appearance is as if the rock was once in a pasty state, and, when so, these rounded bodies were mixed up with it like almonds in cake.

153. **Tendency of Trappean Rocks to the Columnar Form.**—In some cases the tendency of rocks of the trap family to take on a columnar arrangement is shown in the most decided manner. One of the most noted examples is in Fingal's Cave, on the island of Staffa, represented in Fig. 26 (p. 76). Another is the Giant's Causeway, in Ireland. The columns vary from 20 to 200 feet in height, and are jointed, presenting, therefore, an appearance as if they had been built up by putting one pentagonal or five-sided stone upon another. Where the sea has dashed against them they are more or less worn

Fig. 26.

away above, and below extend to unknown depths below the water. Looking like ruins of some ancient work too great for man, it is not strange that popular tradition has connected with them the agency of giants. In most cases the tendency is only partially exhibited, as, for instance, in East and West Rock, at New Haven. The columns are prisms, having from three to eight sides, commonly five or six, and they are generally divided into blocks by joints, as seen in Fig. 27. Each block is usually concave on its upper part, having the lower convex end of the block above fitting into it. Commonly the columns stand

Fig. 27.

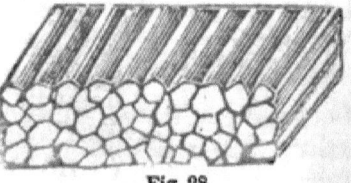

Fig. 28.

upright, but sometimes they are horizontal, as in Fig. 28, representing a basaltic dike in North Carolina. The appearance in this case is that of a wall built as a fortification. Sometimes the columns are in the position represented in Fig. 29. Sometimes the columnar and massive forms are conjoined, as seen in this figure. A remarkable example both of this position of columns

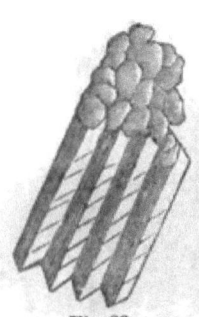

Fig. 29.

and their union with the massive rock is to be seen on Mount Holyoke, in Massachusetts, in Titan's Piazza, so called because there is a group of columns of greenstone pointing downward from the overhanging rock above.

The manner in which rocks of this family are formed will be considered in another part of this book.

154. **Trappean Rocks in this Country.**—Beginning at the north, there is a belt of trap extending 130 miles along the Bay of Fundy, where the violence of the waves has exposed to view magnificent groups of columns three and four hundred feet in height. In the neighborhood of Boston, at Nahant, Lynn, etc., there are ridges, some of them rising to the height of 500 feet. A range beginning at East and West Rock, New Haven, extends up the valley of the Connecticut almost to Vermont, including Mounts Tom and Holyoke, which rise to a height of over a thousand feet. The noted Palisades, on the Hudson, are greenstone. Three ridges of trap extend through the State of New-Jersey, and trap rocks are found in beds and elevations as far south as North Carolina. But it is west of the Rocky Mountains that trap most abounds and reaches the highest elevations. Columbia River has on each side of it mountains of trap, in some cases even a thousand feet in height. The appearance of the columns is seen in Fig. 30, p. 78.

155. **Lavas.**—The materials thrown out from volcanoes become, as they solidify, what are called lavas. The various kinds differ in structure and composition, and, of course, in color. There are two classes of lavas—the light colored, in which feldspar is the chief constituent, and the dark colored, or basaltic, which are grayish-blue, even to black. The structure of lavas depends upon circumstances. That which cools under pressure, and while shut in from the atmosphere, is compact; but

Fig. 30.

that which is at the surface, owing to rapidity of cooling and the action of the air, is porous, and has the name of *scoria*. What is called *pumice*, it is supposed, was cooled by being thrown into water. Though its mineral constitution is the same with the hard, compact lava, it will float on water from its great porousness. It is used for polishing various substances, as wood, ivory, marble, glass, parchment, etc. *Obsidian* and *pitchstone* are vitreous or glassy lavas, the former resembling glass more than the latter, which receives its name from its pitchy lustre. The lavas are very much like the rocks of the trap family, a fact which is of much significance in relation to the formation of the latter, as you will see in a future stage of our investigation.

There are other rocks—limestone, serpentine, etc.; but these I have spoken of sufficiently before.

## CHAPTER IX.

### THE EARTH AS IT IS.

**156. The Earth as a Whole.**—The earth is a ball nearly round, composed of solid, liquid, and gaseous substances, the gaseous occupying the interstices of both solids and liquids, there being also a gaseous envelope around the earth of about fifty miles in thickness. The most abundant liquid is water, which fills up all the cavities on the earth's surface. If the solid part of the earth were uniform instead of having its present diversified condition, it would every where be covered with water, the atmosphere remaining, as now, outside of the liquid envelope. The earth, as a whole, has been found, by various observations and calculations, to be between five and six times the weight of water, and about two and a half times as heavy as the average of common rocks. This great weight of the earth is owing to the pressure to which its internal parts are subjected from the influence of gravitation. Every thing is attracted toward the centre, and therefore any internal portion of the globe is pressed toward this centre by the weight of all which is outside of it. It is just as the lower portions of the air —that is, those which are close to the earth—are condensed by the pressure of those portions outside of them through gravitation, as illustrated in § 152, Part I. It has been calculated that at the depth of 34 miles in the earth air would be so condensed that it would be as heavy as water, and at the depth of 362 miles water would be so condensed that it would be as heavy as mercury, although here upon the surface of the earth water is so little compressed by the strongest pressure man can bring to bear upon it that it is regarded practically as incompressible. It is also calculated that steel at the centre of the globe would be compressed into a fourth of its bulk, and most rocks into an eighth of their bulk.

**157. Form of the Earth.**—I have said that the earth is *nearly* round. Its deviation from a perfectly spherical shape is very slight—that is, compared with its whole bulk. Though it bulges out at the equator 13¼ miles, this, in a ball eight thousand miles in diameter, is but a small matter. This can be shown by the diagram, Fig. 31. Let the curved line PEPE represent the circum-

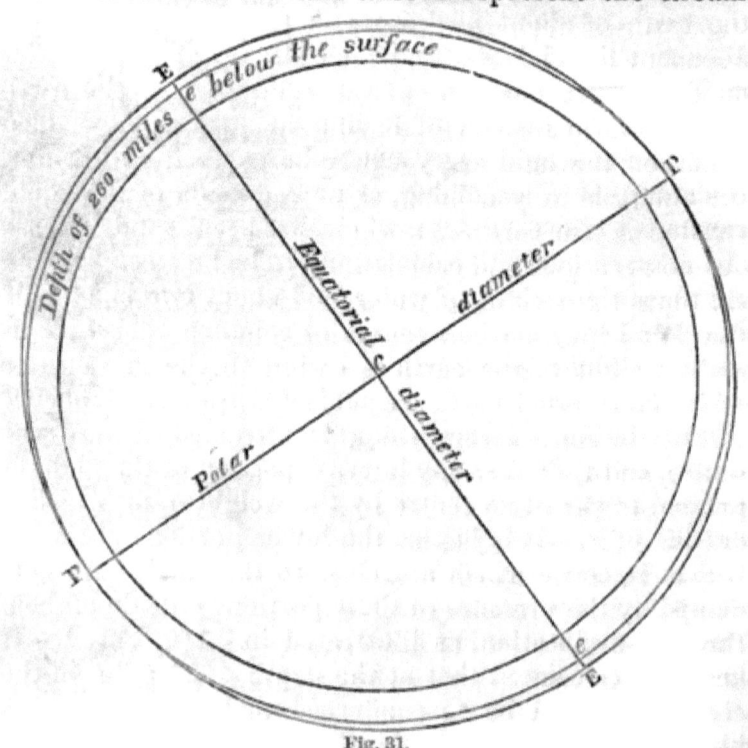

Fig. 31.

ference of the earth as it is with the line running through the poles, PP. But this is not a true circle, for the diameter EE is longer than the diameter PP. The true circle described by the diameter PP revolving on the centre $c$ would be $PePe$. The truth in regard to the shape of the earth is represented here, but much exaggerated—how much you shall see. The line PP is on the scale of 2600 miles to the inch, to represent the 8000 miles of the earth's diameter. Of course $\frac{1}{10}$th of an inch would be 260 miles, and to make the amount of exaggeration alluded to above apparent to the eye, an inner circle at this distance from the outer one is drawn. You see, then, that a tenth of the distance between the outer and inner circle ($\frac{260}{10}$) would be 26 miles, and the bulging of the earth at the equator is but half of this amount. To get at the real, unexaggerated fact, then, you must take only the $\frac{1}{20}$th part of the distance from $e$ to the inner circle to represent the bulging of the earth, and this would be adequately represented by merely doubling the line of the figure at $e$, and letting it gradually diminish to the single line as you go to the poles PP. But small, comparatively, as this equatorial bulging is, it is supposed to be of very great importance in tending to keep the earth always revolving uniformly on one axis—the polar axis. With the very rapid revolution of the earth (1000 miles every hour), and its more rapid motion in its orbit (68,000 miles every hour), even a slight deviation from regularity in the revolution would be attended with disastrous results. There are inferences to be drawn from the shape of the earth in reference to its formation, but these will be spoken of in another chapter.

158. **Crust of the Earth.**—In the crust of the earth, so often spoken of by geologists, is included all that portion of the globe of which we have any knowledge through their investigations. Of course we know nothing definitely and certainly of the contents of the interior of the earth; but great depths have been reached in mining, and besides, as you will see in future portions of this book, there have been upheavals of the crust which have opened to view vastly greater depths than the miner has ever reached. The foundation of this crust is every

where rock. However deep may be the cavities which hold the water, there is a rocky bottom, and, dig wherever you will, rock is found beneath the earth and sand. It is said by Bakewell that "the crust of the globe with which we are acquainted does not exceed, in comparative thickness, that of a wafer to an artificial globe **three feet in diameter.**"

159. **Land and Water.**—The surface of the globe is reckoned as containing about 211,000,000 square miles, of which about 150,000,000 are covered with water, and 61,000,000 appear as land. The proportion of water-surface to land-surface is nearly 8 to 3.

160. **Elevations and Depressions in the Earth's Surface.**—The elevations in the forms of hills, table-lands, and mountains vary exceedingly, the highest being Mount Everest, in the Himalaya range, which is 29,000 feet, or five and a half miles high. The ocean varies much in depth, sometimes probably reaching to 50,000 feet, though no satisfactory soundings have ever ascertained such a depth. The average depth is somewhere from 15,000 to 20,000 feet. Often about the continents there is a fringe, as we may term it, of shallows. On the coast of the United States, off the State of New Jersey, there is such a fringe 80 miles in width, the depth at its edge (where the ocean really begins) being only 600 feet, while the depth across the ocean, in a line from Newfoundland to Ireland, has been found to be from 10,000 to 15,000 feet. Some extensive waters, also, between different bodies of land, can not be regarded as parts of the ocean. In the waters separating Great Britain from Europe the depth is less than 600 feet, and in a large part of the German Ocean it is only 93 feet. Similar facts have been discovered in regard to the waters between Asia and the islands appended to it. These should really be considered as a part of the continent of Asia, and Great Britain as a part of the continent of Europe.

Although the projection of the earth at the equator is

so small compared with the whole earth, as was shown by Fig. 31, yet it is much greater than the mountainous projections on its surface, for the amount of this projection is $13\frac{1}{4}$ miles, while the height of the highest of the mountains is $5\frac{1}{2}$ miles. It is to be observed, also, that the depressions are much greater than the elevations, with the exception of the grand equatorial one, for in some cases they probably reach, as already stated, a depth of 50,000 feet; and, if the ocean were laid bare, we should see irregularities like those on the land, but on a much larger scale. Between the so-called Banks of Newfoundland and Newfoundland itself there is one of those deep submarine valleys, while at the Banks there is a plateau of rock that comes within about 250 feet of the surface. The water on this plateau abounds in fish, which choose such shallow places rather than the deep ocean valleys. South of the Banks the Atlantic Ocean reaches the immense depth of 30,000 feet.

**161. Arrangement of the Land.**—There are some peculiarities in the arrangement of the land of the earth that are worthy of notice. The great bulk of it is in the northern hemisphere, up about the arctic region, there being nearly three times as much land north of the equator as there is south of it. As it extends from the north down into the southern hemisphere it narrows very much, as seen in the shape of North America and Africa, which is triangular. The same disposition of the land is essentially carried out in Hindostan and other extensions southward of the continent of Asia. Notice, besides this, that the land is really divided into two great parcels, the Eastern and Western Continents, for Europe, Asia, and Africa may be regarded as one; and that there are two great oceans between them, one of them being much larger than the other. It may be remarked here that, in the present state of the world, it is a very great advantage and convenience to have the narrower ocean, the Atlantic, lying between those portions of the world which, from their common civilization, have so much intercourse with each other. It would be a great bar to this intercourse to have so wide an ocean as the Pacific lying between America and Europe. I have spoken of the two grand masses of land as extending from the arctic region down into the southern hemisphere; but observe that it is not an unbroken extension of land. By an arrangement of the waters, as noticed by Professor Guyot, there is a band of water from east to west cutting the land in two almost completely—so nearly that it requires only two short canals to be made by man to finish the communication. This belt is composed of the two oceans, the **Mexican Ocean in the** western hemisphere, **and the Mediterranean Sea,** the Red Sea, and the seas of the East Indies in the eastern. You can readily see how this arrangement promotes largely the intercourse by water throughout the earth, and how important it is that those parts of the plan which the Creator has purposely left to the ingenuity and enterprise of man should be carried into execution.

162. **Arrangement of Mountains.**—Mountains are not scattered about here and there in a confused manner, but there is an obvious general plan in their arrangement. They are to a great extent placed in chains or ranges, as you see in the Rocky Mountains, the Andes, the Appalachians, the Pyrenees, the Urals, the Himalayas, etc. Perhaps the most remarkable fact about these mountain chains is that they are so arranged as to inclose large basins of land on the continents. Thus in North America the immense basin watered by the Mississippi and other rivers lies between the great ranges of mountains on the east and the west. It is this basin arrangement, with great mountain walls, that is the **grand** characteristic **of a continent in** distinction from an island. Notice another great fact in the arrangement. The highest mountains are placed on that side of the continent which is toward the broadest ocean.

Thus the Rocky Mountains are higher than the range of mountains which run down from Maine to Georgia, called by the general name of the Appalachian range. So also, as the Pacific Ocean is broader opposite to South America than opposite North America, the Andes are much higher than the Rocky Mountains. The same plan, essentially, is carried out in Africa, Australia, and also in Europe and Asia, though not in so clear and simple a manner.

There is much variation in the arrangement of different ranges of mountains. They are more commonly curved than straight, and the manner in which the individual mountains stand toward each other is very different in different cases.

There are certain prevalent directions of the chains of mountains, and the same is true of groups of islands, which are often really the summits of submerged mountain ranges. In the Pacific Ocean they have usually a northwesterly direction, or trend, as it is termed.

163. **Volcanoes.**—Some mountains throw out occasionally from their summits large quantities of liquid lava and matter in gaseous form; often also solid rocks, ashes, mud, etc. The exhibitions made in these eruptions are among the most magnificent in nature. About two thirds of the volcanoes are on islands, and the greater part of the other third are near the ocean on the continents. This nearness to water is supposed to show that steam must have much to do with the eruptions, which also seems to be proved by the fact that in these eruptions great quantities of steam escape. In Fig. 32 (p. 86), in which the shaded parts indicate the localities of volcanoes, not only is their vicinity to the ocean shown, but the fact that, like ordinary mountains, they are very often arranged in ranges.

164. **Plateaus and Lowlands.**—Any extended region, whether it be flat or diversified, if it be much elevated above the level of the sea, is called a plateau, while the

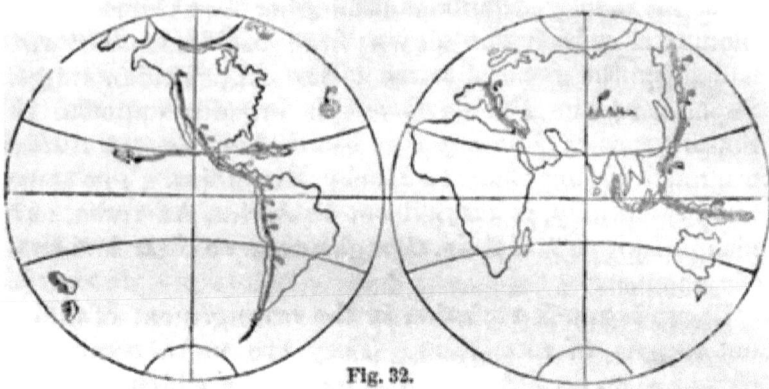

Fig. 32.

term lowland is applied to regions that are less than a thousand feet above that level. The great valley of the Mississippi, the plains of the Amazon, and the pampas of La Plata, are examples of the latter. Plateaus sometimes extend between ranges. Both plateaus and lowlands may have single mountains or mountain ridges. The term table-land is often applied to plateaus, especially when their surface is little diversified. One of the most elevated plateaus in the world is that of Thibet, lying between the Himalayas and the Kuen-Luen Mountains, the height varying from 11,000 to 15,000 feet. The State of New York is a plateau, varying in height from 1500 to 2500 feet. The Great Salt Lake lies in a corner of a plateau by the Rocky Mountain range, which has an altitude of from 4000 to 5000 feet. There are some lofty plateaus in South America. The city of Quito is situated on a plateau at an elevation of nearly 10,000 feet, and the city of Potosi on one at an elevation of 13,330 feet, the Lake Titicaca being on the same plateau at the height of 12,830 feet. We often see the same arrangement in a small way among hills of ordinary size. An elevated plain skirts a hill, or lies between hills, making a plateau of small extent. There may be a pond also in such a plain, like the lakes that are sometimes found in the immense plateaus.

165. **Rivers.**—Rivers come from mountains and plateaus, and, having gathered the waters from these, run through the lowlands into the sea. There is system here as well as in the arrangement of mountains. For example, the great basin included between the Rocky Mountains and the Apalachian range has a grand river system, the Mississippi being its principal river. Then there are two other great systems in North America, in which the St. Lawrence and the Mackenzie are the chief rivers; but these are not as extensive as that of the grand basin of the continent. In South America there are also three principal systems, that of the La Plata, corresponding with that of the Mississippi in North America, that of the Amazon in the east, and that of the Orinoco in the north. The systematic arrangement, which is so obvious in the case of the large rivers of the globe, exists as really with the smaller rivers, though, from the influence of local circumstances, it is not always as manifest.

166. **Lakes.**—Lakes are formed by rivers and streams of various sizes, which pour their waters into depressions on the land from which they can not escape readily, in some cases not at all. These depressions are of various shapes and sizes. Those from which the water has no outlet are salt lakes, as, for example, the Dead Sea in Asia, and the Great Salt Lake of this country.

167. **Relation of Mountains to Fertility.**—The position of mountains has a great influence upon the fertility of a country by detaining the winds and condensing their moisture. Observe how the winds are produced. There are two principal causes of them, heat and the rotation of the earth, and consequently the winds that blow from the tropics, or trades, as they are called, are eastern winds, while the winds coming from the colder regions are western. The winds from the east, being warm, are therefore loaded with moisture, while the cold western winds have comparatively little moisture, and continual-

ly gather or take up more as they bend toward the equator, and are therefore drying winds. For this reason, more rain falls on the eastern side of a continent than on the western. The annual amount of rain in Europe is 32 inches, while in the temperate zone in the United States, east of the Mississippi, it is 44 inches, the difference being still greater if we compare the eastern part of South America with the opposite side of the Atlantic Ocean. You can see now what influence the mountains have on fertility in the case of the American continent. Its highest mountains are on the western side, and therefore the warm eastern winds sweep with their moisture over almost the whole breadth of the continent, parting with it here and there from condensing influences. Fertility is the result of this diffusion over the continent of this moisture, and hence America is appropriately called by Professor Guyot the Forest Continent. Where the moisture deposited from the passing wind is not enough to produce forests, we have the prairies and the pampas. Now if the low Appalachians were on the west and the high Rocky Mountains were on the east of North America, most of the moisture of the trades would be condensed at once, and be poured back into the Atlantic Ocean, and in that case the great river system of the continent's basin would not exist, but in place thereof a desert. You can readily apply for yourself the above explanation to the desert of Sahara, in Africa, and other great deserts, and therefore I need not dwell farther on this interesting subject.

168. **Circulation of Water.**—Water is every where in motion on the earth. Rising continually from every point of the surface by being dissolved in the air, it is brought down again to the earth by being condensed from its vaporous state. There is, therefore, a constant circulation of the water back and forth between the earth and its envelope of air. In the earth itself it is never at rest, but insinuates itself among all loose parti-

cles and masses, runs down all declivities in streams small and large toward the ocean, and is agitated by the winds and the tides. Not only is its motion kept up in these ways, but there are great systematic currents in the ocean, which maintain a free circulation between different quarters of the globe, and exert a marked influence upon the climates of many countries. The great Gulf Stream is one of these currents. By all of these means of motion there is secured a circulation of water in the earth which in its system and thoroughness bears an analogy to the circulation of the blood in the body. What extensive and varied effects it produces in this circulation upon the solid materials of the earth's surface you will learn in succeeding chapters.

169. **How the Earth's Surface is Diversified.**—I have already spoken of some of the grand features of the earth's surface. But there are subordinate features which diversify it greatly—bluffs, hills of various contours, valleys with and without streams, collections of rock differing in arrangement, shape, and color, some being stratified and others not; rocks jutting out from the ground, boulders of various sizes, etc. A material addition is made to this diversity by water every where enlivening the scene by its motion in obedience to every impulse. And then we have life, with its endless variety of shape, and color, and motion.

170. **Treasures in the Crust of the Earth.**—As you have already seen, there are precious treasures of every variety scattered among the rocks, and earth, and sand of the globe for the use of man. There are immense stores of coal; metals of every kind, and in quantities suited to the amount of uses to which they can be appropriated; precious gems of every quality and hue; granite, sandstone, marble, etc., for building; clay for pottery and bricks, etc.

# CHAPTER X.

### PRESENT CHANGES IN THE EARTH.

**171. Ages of the Earth.**—The earth is older than 6000 years, as has been absolutely demonstrated by the researches of geologists. Here is an apparent discrepancy between what is revealed by the works of God and his written revelation; but it is only apparent. I shall speak of this point particularly in another place, and let it now suffice to say that it is the belief of most geologists that the days of Moses represent long ages, and that the time which has elapsed since the earth was finally fitted for man is a very small period compared with the length of time expended by the Creator in its construction and preparation. It is my intention, in this chapter, to treat of those changes which are matter of observation and history. The word *present*, in the title of the chapter, refers to the whole time in which man has inhabited the earth, in distinction from the long periods consumed in its preparation for his use. It is not always easy to make this distinction, and some changes will be noticed in this chapter which began far back of the age of man. This is necessary, because the same agents which prepared the world for its present purpose have produced all the changes since man was introduced upon the scene.

**172. Agents of Change.**—The agents by which changes are constantly produced in the earth are heat, water, air, chemical processes, attraction, electricity, and life, both vegetable and animal. The manner in which they act will be developed as we proceed, and therefore I will not dwell on this subject here, but will only notice in a few words the principal of these agents. Water and air are every where kept in motion by the influence of heat and attraction, as was fully explained in Part I. But water has a larger agency in producing the changes in the earth than air. Through the influence of heat it is carried up constantly into the atmosphere in evaporation, and then falls in rain, snow, etc., and in seeking its level in obedience to attraction, effects a large portion of these changes. The influence of heat is seen in the eruptions of volcanoes and other phenomena, and these indicate to some extent the agency which heat had in building up the earth, as do the present effects of moving water what agency that had in the work. It is these aqueous and igneous agencies, as they are termed, which, often acting in opposition and sometimes in unison, have for the most part arranged and consolidated the materials of the earth, so as to put it into its present condition; and therefore it is necessary to view their present operations, that we may understand those which were carried on in the ages that preceded the advent of man, when the earth was being prepared by successive steps to be his habitation. This will be, then, the line of our investigation in the present chapter.

173. **Water Changing the Locality of Materials.**—What we see in the washings of every shower on slopes and hill-sides exemplifies some of the vast changes which water is producing on a large scale in the earth. Different kinds of materials are moved according to the different degrees of rapidity with which the water flows. If you look at a mountain torrent you see nothing in its course but large stones, because not only mud and sand, but pebbles of considerable size, are carried along by the force of the water. If you go farther on, where the stream is less rapid, you will find the pebbles, and if the rapidity lessens as you follow the stream, there will be sand on its bottom; and when the water moves on slowly through a plain, you will find the sediment deposited to be mud. The explanation of this sorting out of material by moving water has been given in § 193, Part I.,

and it is not necessary to repeat it here. The pebbles, and sand, and mud moved by water were all once in solid rock, and how the rocks are thus broken up will be seen hereafter. The deposit of these materials is regulated by other varying circumstances as well as those which I have mentioned, as I will now proceed to illustrate.

**174. Deposits in Rivers and on their Borders.**—A river descending some slope of considerable pitch always brings along with it much solid matter, and if it afterward move sluggishly through a plain, this matter will be deposited quite evenly over its bottom. If the river be narrowed much at any point the water will move so rapidly there that it will carry the material suspended in it on beyond, and then, if it spread out suddenly over a large surface, there will be much matter deposited, except in the direct line which the main body of the water takes. On either side of this channel—that is, on the flats—will the most of the deposit be made. In some cases the plain on either side of a river is so high that the water flows over it only when it is temporarily raised by a freshet. This is termed the flood-plain. In this case there is some addition to the plain every time that a freshet occurs, until at length the land is raised so high that the flood does not overflow it. This process, year after year, tends to make that part of the plain immediately bordering on the river higher than the rest of the plain, because, the farther the water is from the river, the less has it suspended in it from the settling of the sediment which is continually going on. And as some of the matter is deposited on the bottom of the river itself, the river is gradually raised, and may be, after a while, on a level which is above the adjoining land, except that which forms the banks. These results are shown by Figs. 33 and 34. Let $a$ $a$, Fig. 33, be a flat valley, and $b$ a stream of water which runs

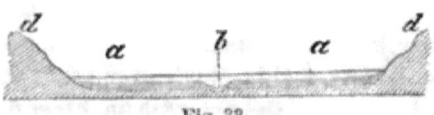
Fig. 33.

through it, occasionally overflowing its banks. In the overflows mud will be deposited over the valley-flat, and

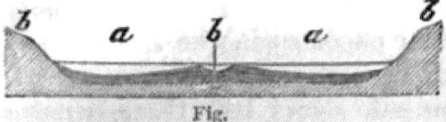
Fig.

after a while the result indicated in Fig. 34 will be realized. Just this process has been gone through with along the Mississippi, so that there is a natural levee, as it is termed, on each side of the river. The inhabitants are obliged to raise this in certain places to prevent inundations from freshets. Sometimes a break (crevasse) occurs in the levee, and much damage is done in the country adjoining. The same process goes on, and the same results occur at length in those cases where the land along a river is inundated at all seasons.

175. **Improvement of Rivers.**—A very pretty illustration of the manner in which the solid material in rivers is deposited is furnished us by a mode of improving river navigation sometimes practiced. We will take a particular case. The river issues from a gorge, and then spreads out over a large flat surface, there being, of course, a channel through the flat for the main body of the water to flow in. The object of the plan of improvement is to prevent accumulation of sediment in the channel, and even, if possible, to carry out sediment already deposited. This is done by building piers of stone across the flats out toward the channel, as indicated in Fig. 35.

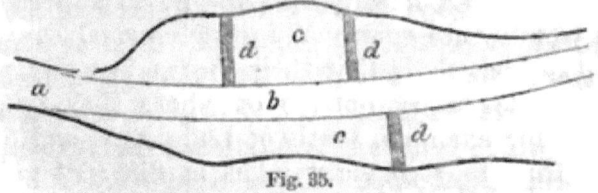
Fig. 35.

Let $a$ be the gorge, $c\ c$ the flats across which the piers stretch, as represented, and $b$ the channel. The piers, $d$, $d$, $d$, it is obvious, will somewhat detain the water on the flats, and so allow the sediment to fall from it in larger amount than it otherwise would; and this detention at the sides of the channel will cause more water to flow through it, and to flow more rapidly. Then, as the flats fill up with sediment, this effect upon the flow in the channel will continually increase.

176. **Deposits in Lakes.**—In every **lake there is** more or less deposit of sediment from the river or rivers that flow into it, and also from all smaller streams that do so, even to the little rill temporarily made by a shower. The tendency, then, is to fill up the lake, and in time the result would be the conversion of the lake into a river. **If a lake have a river** entering at one end and issuing at the other, the deposit is made in that part of it where the river enters, and the extent to which it reaches depends on circumstances. The more rapid the entering river, and the more shallow the lake, the greater will be the space over which the deposit will be made. The water of the entering river is of course turbid, while that of the issuing river is clear, the sediment being all deposited long before the water reaches that end of the lake. In the Lake of Geneva, which is thirty-seven miles long, and varies in breadth from two to eight miles, we have all this exemplified. The Rhone enters it muddy, but it issues beautifully clear at the city of Geneva. The rapidity with which the filling up goes on may be judged of by the fact that the town of Port Vellais, which, eight centuries ago, was on the edge of the lake, is now more than a mile and a half distant from it. The great lakes of this country are continually growing smaller. There are facts which show that their shores, in many places, were once far away outside of where they now are. There is, for example, south of Lake Erie, **and** distant from it from four to eight miles at **different** points, a ridge **made up of sand,** gravel, and rounded pebbles, just as the **shore of the lake is now.** Moreover, when wells are dug, or any excavations made in this ridge, there are found deeply buried in the soil pieces of decayed wood,

small and large, and such shells as are now met with in the lake.

**177. Deltas.**—The name delta is commonly given to the accumulation from a river at its mouth as it empties into a sea or the ocean, when it has gone so far as to form flats, through which the river runs usually with a network of channels. The name is given from the common resemblance in the form of this accumulation to the Greek letter Δ, which is called Delta. The shape which the deposit tends to take is fairly represented by Fig. 36. From

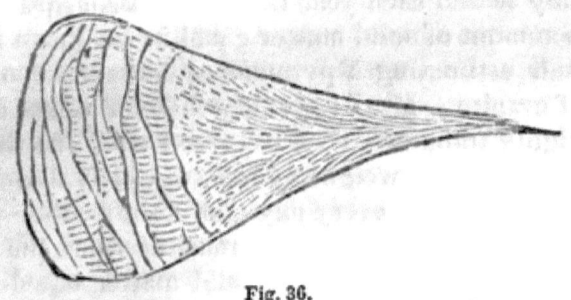

Fig. 36.

the point where the river enters the sea the sedimentary matter spreads out or radiates, and at the same time it is tossed about and beaten back by tides and the waves. It is this action of the sea, in opposition to the prolongation of the river current into it, that produces the bars so often found in such cases. These bars are indicated in the figure at the sea-extremity of the delta. The numerous sand-bars formed at the mouth of the Mississippi render the entrance to it difficult and hazardous. Sometimes even islands are formed. Thus, about fifty years ago, an island was formed opposite the mouth of the Hoogly River, which in 1818 was two miles long and half a mile wide, was covered with vegetation, and was inhabited; but afterward this was gradually swept away, and dwindled to a mere small sand-bank. It is common there, and in other cases also, for islands to form, and then be destroyed by the change of currents from the formation of new islands. There is no delta formed at the mouth of the Amazon, because there is an ocean current that sweeps by, and this carries the sediment discharged by that river up by the coast of Guiana, where it is deposited, forming immense muddy shoals and swampy tracts.

178. **Amounts of Deposits from Rivers.**—The amount of sediment deposited by large rivers is enormous. The Ganges pours such a quantity of mud and sand into the Bay of Bengal that the water is seen to be colored by it sixty miles from the shore. It is calculated that in the rainy season each year this river discharges into the sea an amount of solid matter equal in weight to fifty-six pyramids, estimating a pyramid to contain 6,000,000 of tons of granite. Mr. Lyell states that "if a fleet of more than eighty Indiamen, each freighted with one thousand four hundred tons weight of mud, were to sail down the river every hour of every day and night for four months continually, they would only transmit from the higher country to the sea a mass of solid matter equal to that borne down by the Ganges in the flood season, as the exertions of a fleet of about two thousand such vessels going down daily with the same burden, and discharging it into the gulf, would be no more than equivalent to the operations of the great river. Yet, in addition to this, it is probable that the Burrampooter conveys annually as much solid matter to the sea as the Ganges." The amount of land made by such enormous quantities of sediment is very great. Most of the lower part of Louisiana was formed by sediment brought down by the Mississippi, and the land has encroached upon the water several leagues since New Orleans was built. The solid matter annually discharged by the Mississippi is two thousand million (2,000,000,000) tons. This is sufficient to cover a township six miles square with earth thirty feet deep.

179. **Extent of Deltas.**—The deltas of very large rivers are of great extent. That of the Nile is about as

large as the State of Vermont. That of the Ganges is much larger, being 220 miles long, and having a base on the sea of 200 miles. It is bounded on either side by an arm of the Ganges, and near the sea it is intersected by a net-work of rivers and creeks, and is a resort for crocodiles and tigers. The material constantly brought down the river is encroaching upon the sea, and it now forms a slope extending out about a hundred miles. This really ought to be taken into the account in estimating the extent of the formation.

180. **Consolidation of Deposits.**—In some cases the consolidation into rock of the deposits of sediment by water is going on at the present time, especially where there is mingled with the sand and mud some agglutinating material like carbonate of lime. From this cause there is rock continually forming in the sediment discharged by the River Rhone. In the Museum at Montpellier there is a cannon which was taken from the sea near the Rhone's mouth incased in a crystalline calcareous matter, and having scattered through it broken shells. But the whole subject of the consolidation of mud and sand into rock will be treated of hereafter, and I will not dwell upon it farther here.

181. **Water Encroaching upon Land.**—Thus far I have spoken only of the encroachment of land upon water by the deposit of material brought to lakes and the sea by rivers. But there is sometimes the opposite effect—the wearing away of land by the action of water. Much of the eastern coast of England is wearing away, and many localities of towns have disappeared in the German Ocean. Some of the coast of Long Island is encroached upon rapidly by the sea; and at Cape May, in Delaware, land is destroyed at the average annual rate of nine feet. Sullivan's Island, South Carolina, was worn away in three years to the extent of a quarter of a mile. Some of the Shetland Isles have been destroyed by the sea, and the granite rocks of some of them that are now wasting

away, standing up in the midst of the water, look in the distance like fleets of vessels. On the coast of France, especially in Brittany, where the tides rise to a great height, the sea is constantly encroaching upon the land, and occasionally does so to a large extent. In the ninth century many villages were carried away. Great changes have occurred in Holland from time to time, the land sometimes gaining upon the sea, and sometimes the reverse. At one time the tide, breaking through a dam, overflowed seventy-two villages, and irretrievably destroyed thirty-five of them. Other examples in abundance might be cited, but these will suffice.

182. **Erosive Power of Water.**—Water, acting by itself mechanically, exhibits in the course of time great erosive results. The hardest rocks can not resist it, much less the softer ones. The Pulpit Rock, so called, at Nahant, Mass., seen in Fig. 37, is an example of the erosive

Fig. 37.

action of the waves continually dashing against it year after year. But little of the rock is worn away each day, but this little daily work sums up largely when continued through centuries. Water does much erosive work by rubbing one solid surface against another. This we see on the sea-shore, as the waves, lashing the shore, jostle the pebbles, great and small, against each other. The same thing is done with the grains of sand and mud wherever there is water moving them. Every pebble and grain that is carried down by a river toward the sea grows continually smaller by being rubbed on its passage by the accompanying grains and pebbles, as well as by the friction of the water itself. We see this mode of erosion exemplified in the pot-holes seen in rocks where there is shallow water running over them. The stones contained in them are rounded by the constant friction, and at the same time wear the hole in the rock continually larger. In the Franconia Notch of the White Mountains there is a pot-hole in granite, called the "Basin," which is fifteen feet deep and about twenty in diameter.

183. **Niagara Falls.**—One of the most striking examples of the erosive power of water we have in the Falls of Niagara. The river runs over hard limestone, but under this are soft shales. The result is that the water at the Falls continually wears away the shales, and the limestone falls from want of support below, the water above pressing it downward. The edge of the fall is therefore constantly receding at a rate calculated variously by different persons, from one foot to three feet annually. This is a slow recession, but in the course of centuries the change is a great one. There is the most decided proof that the Falls were once seven miles farther down the river than they are now, so that they have receded all that distance, the process having begun long ages before the creation of man. As the layers of rock, instead of being perfectly horizontal, dip a little in running back toward Lake Erie, the height of the fall is constantly lessening as it recedes. It can therefore nev-

er recede entirely to the lake, because, before it reaches there, the fall will cease to be sufficient for the wearing away of the shales underlying the limestone.

**184. Cañons of Colorado.**—The case just cited of erosion by water, though so striking, is by no means so strong a one as some others; for, while the rocks eroded by the water of the fall are soft, there are instances of vast and rapid erosion in solid tough rock. The strongest example is in the cañons of Colorado, which are rivers running between perpendicular walls of rock, as represented in Fig. 38, in some cases standing even 6000 feet in height. These passages through the rocks were actually worn by the water, incredible as it may at first thought appear. We have some comparatively recent erosions, which show that such monstrous erosions are possible, if there be a sufficient length of time allowed for them. For example, the River Simeto, having been dammed up by an eruption from Mount Etna in 1603, cut a passage through hard blue basaltic rock in a little over two centuries, which was from fifty to several hundred feet in width, and in some places fifty feet deep. It is rather difficult to realize that water can accomplish such an amount of erosion, but a little reflection on some common results with which we are familiar will help us to do this. We see the stone steps of public buildings very soon worn by the friction of the many feet that pass over them. But this friction is not constant; it is so intermittent that it occupies but a small portion of the time that passes from day to day. If it were constant, the stone would soon be so much worn as to require being replaced by another. Now the friction of water erodes stone like the friction of footfalls; and if it be constant through century after century—perhaps age after age—the erosion, we can see, will be great in amount.

**185. Rocks Disintegrated by Frost.**—As there are crevices and interstices in rocks, the water enters into these, and in cold weather becomes frozen there. In freezing,

PRESENT CHANGES IN THE EARTH. 101

Fig. 38.

as you learned in § 329, Part I., it increases in bulk. The result of this expansion of the water is that it tears the rocks in pieces, the fragments being of all sizes, from grains up to even large masses. The softer rocks, as the shales, are broken into small fragments as the water gets into their numerous small interstices; but in the case of the harder rocks, the water enters cracks and crevices that it finds here and there, and the fragments separated by the frost are of considerable size. It is chiefly from this action of frost on rocks that, at the foot of such rocky fronts as are presented by East and West Rock at New Haven, and by the Palisades of the Hudson River, there are accumulations of fragments of rock of various sizes. Such an accumulation is called a *talus*. In Fig. 39 one of these is represented. You see a difference in size of the stones in different parts of this talus. This results from the fact that every agitation from

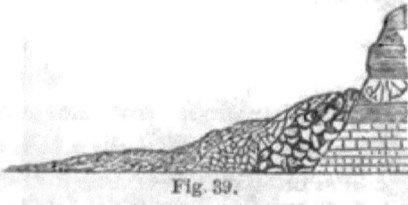

Fig. 39.

the falling of the fragments, and the action of water and wind, tends, by friction, to make the fragments smaller, and to remove the smallest of them away from the foot of the rock. Some of them are, in process of time, converted into sand, and even fine powder, and these the rains may remove, or the tides, if the foot of the rock be so situated as to be exposed to them.

186. **Chemical Action of Water.**—Water ordinarily contains dissolved in it both solids and gases, which enable it to act chemically upon the rocks. For example, the limestone or calcareous rocks are acted upon by water containing carbonic acid gas in it; and as rain, falling through the air, becomes charged with this gas, which it finds there, water is continually carrying off calcareous matter from such rocks, supplying thus the sea with material for the shells of the animals that live in it in such abundance with these coverings upon them. Water thus charged also acts with facility upon the oxyds and sulphurets of iron which are so often present in rocks, and also upon the alkalies that enter into the composition of

the feldspathic rocks, so that even the harder rocks are not full proof against the chemical action of water. It is from this action on these rocks that the silex is provided for the grains, and grasses, and other plants, in a condition to be dissolved and carried up the plant in the sap. Even the granite, of which man constructs his most enduring monuments, is every where being slowly worn away by this action, and thus ever furnishes its quota of fertility to the soil in the alkali and clay that come from its decomposing feldspar.

187. **Weathering.**—This is a term which includes, to some extent, the mechanical action of water with its chemical. The idea is that rocks, on exposure to the weather—that is, to rain and air, are more or less disintegrated, and sometimes even changed in chemical constitution, either at the same time or after the disintegration. This weathering is sometimes exhibited in a very striking manner by ancient monuments. This is the case with the Druidical monuments in the north of England, which are constructed of a very hard rock called millstone grit. Three monstrous pillars, called the Devil's Arrows, having had the rain beating upon them for two thousand years, are furrowed deeply all down their sides, the furrows being deepest at their summits, and becoming wider and less distinct toward the bottom. Crags of rock too hard to crumble under the weathering often present the same furrows. Where rocks are porous, and, as we may say, loosely constructed, as are many sandstones, the disintegration of weathering goes on largely and rapidly, and we have in such localities imperceptible gradations from earth to solid rock. This may be seen often where the rocks jut out from the soil, but more especially in digging down through the successive gradations of soil and crumbling rock to the hard rock itself. The same weathering which disintegrates the rocks does the same work among the fragments of rock, small and great, in the soil, for both water and air

are present there. Sometimes stones scale off from this weathering process. In many cases this occurs in a very regular manner. A beautiful example of this I found in a boulder of greenstone nearly two feet in diameter. It had a covering, in the form of a shell, of about a third of an inch in thickness, of the same composition with the boulder itself. Parts of this shell had been broken off. but it was evident that it had been whole quite recently, The first thought of any one that had no knowledge of such results of weathering would be, that the boulder had in some way received a coating of stone; but this could not in any way occur, and we are shut up to the conclusion that this rind, as we may call it, of the boulder was separated from the body of the stone by the weathering process.

It is supposed that the famous Loggan, or rocking stones, referred to in § 100, Part I., have been shaped so as to rock by this weathering. If a rock having a small base rest upon another, the weathering will be apt to go on quite rapidly on the under side, from the dampness and the continued shade, and the base may after a while become so narrow that the rock, though a large one, may be easily moved back and forth upon it.

188. **Glaciers.**—Having spoken of the effects of water in its liquid state, I pass now to consider those which are produced by it in its solid condition. You will find in another chapter that, great as these effects are at the present time, they were vastly greater in ages of the world long gone by. Both glaciers and icebergs had a large agency in preparing some portions of the earth for the use of man.

A glacier is simply a river of ice. It flows down a valley from the regions of eternal snow and frost above. It does not merely slide, for the ice is somewhat plastic, and so bends to accommodate itself to the different widths of the valley and to uneven surfaces. The flow, it is true, is very slow, but it is constant; else the snow that feeds it above would increase from year to year. The rate of flow varies from varying circumstances. Professor Hughes built a house on a glacier, and he found it to move during fifteen years at the average rate of eight inches daily, or over three quarters of a mile during the whole time. A glacier is made out of snow which becomes consolidated into ice, partly by pressure and partly by the infiltration of water, which, by freezing, unites all the grains of snow together. In Fig. 40 (p. 106) is represented one of the grand glaciers of the Swiss Alps, the glacier of the Viesch. A glacier carries along whatever of loose material it finds in its course, and therefore there is always a row of loose stones, of various size, lying along upon the ice on each side of the glacier. These are called the lateral *moraines*. In the glacier of the Viesch there is, as you see, a row of stones along the middle. This, which is called a *medial* moraine, arises from the union of two glaciers in one, as two rivers of water unite. In this case the two lateral moraines of the glaciers which are adjacent to each other unite at the confluence.

189. **Termination of a Glacier.**—The colder is the climate the farther down does the glacier extend. The termination is higher up in the summer than in the winter, and it varies in different seasons, according to the temperature of the season. The locality of the end varies sometimes even miles in the course of a series of years. The glacier is sometimes spoken of as retreating, but this language is, of course, not strictly correct, for there can be no movement backward. The apparent retreat is owing to the melting of the lower part of the glacier to a higher point than before. Toward the termination of a glacier the moraines become less and less distinct from the melting of the ice, and at the very end there issues a stream of water, carrying along with it, to a greater or less distance, much of the loose material which the glacier has brought down from the rocks. The stream

Fig. 40.

which comes from the glacier of the Viesch is seen in Fig. 41.

Fig. 41.

190. **Effects of Glaciers.**—Fragments of rock of various sizes, from large stones to pebbles, and even sand, become imbedded in the bottom and on the sides of the glacier. These, held firmly in the ice, rub on the bottom and sides of the valley, and when these are laid bare by the melting of the glacier toward its termination, the rocks exhibit varous marks, as grooves, striæ, scratches, roundings, smoothings, etc., according to the shape and character of the fragments that have been brought in contact with them under the immense pressure of the glacier. Stones that are loose are crushed and ground to sand, some even to mud. Fig. 42 (p. 108) shows the side of a glacier valley after the melting of the glacier. The rocks are striated, smoothed, and rounded wherever the ice has been, while the portions which were above its reach present the usual rough and ragged appearance of hard rocks.

All the crushed and ground matter is at length

Fig. 42.

brought to the termination of the glacier, together with the stones imbedded in the ice and those accumulated on the surface. Much of the finer matter is carried away by the stream of the glacier, to be deposited here and there, while the more bulky parts are heaped up at the end of the glacier in what is called the *terminal moraine*. As the limit of the glacier varies, as already stated, very much from year to year, these terminal deposits are scattered in heaps over quite a large space, and are much changed from time to time by the moving ice and water.

191. **Icebergs.**—In very cold regions the glaciers extend down to the very borders of the sea, and the end of a glacier breaking off into the water, forms an iceberg. Icebergs are of various sizes, many of them reaching a height of 200, some even 300 feet above the surface of the water. As, from the specific gravity of ice, only one twelfth of it is above the surface, we see only a small part of an iceberg. For the 300 feet that we see there are 3300 feet (considerably over half a mile) beneath the surface of the water. A representation of an iceberg seen by Captain Ross is given in Fig. 43. Ice-

PRESENT CHANGES IN THE EARTH. 109

Fig. 43.

bergs are sometimes very extensive. A French expedition measured several which were a mile in breadth, and one which was 13 miles long and 100 feet high. Icebergs appear often in great numbers. Scoresby counted 500 of them starting from the frozen regions at one time for the south. Dr. Kane saw 280 in Baffin's Bay at one time. Like glaciers, icebergs are more or less loaded with fragments of rock, great and small, and this load is dropped in the sea as the iceberg melts. It is supposed that the Banks of Newfoundland were in great part made by deposition from icebergs, and the accumulation is constantly going on. These bergs not only drop material, but they grind much of it up into sand, and even mud, by dragging, as they often do, on the floor of the sea. Sometimes they are stranded, and then, moved by the waves, they roll back and forth, stirring up the muddy bottom, and crushing any fragment of rock that icebergs may have dropped there. Captain Couthoy saw one stranded on the Grand Bank, and the water was charged with mud at the distance of a quarter of a mile.

**192. Water a great Leveler.** — By the various ways which I have mentioned, water is continually making additions to the comminuted solids of the earth, the sands, and the soils. This is going on in a large way by the deposit of sediment by large rivers, by the streams at the terminations of glaciers, and by the grindings of icebergs. But much more of this is effected in the aggregate by what water is doing in a small way in every part of the earth's surface. The weathering that is done all around you every day is the representative of vast operations that are gradually changing matter in form, locality, and properties over the whole globe. The tendency of all this, on the whole, is to bring the heights of the earth down, so that water may be called the great leveler. If there were no operations going on, or to be instituted, in opposition to this tendency, in time the solid matter of the earth would be one level, the oceans being filled up with the ruins of the rocks, the water therefore covering up the land, and making a universal ocean. It would require long ages, it is true, to produce this result, but it would finally come. What operations tend to prevent this will be seen farther on.

**193. Agency of Heat.** — As water generally exerts a leveling influence, heat elevates, and thus tends to preserve the equilibrium in the earth's crust. While water wears down rocks, volcanoes are throwing up melted rock to be consolidated as it cools; and earthquakes, which are connected with volcanic action, lift up, as you will soon see, vast tracts of country to a higher level, while the same elevating process is going on in other tracts in a gradual manner. The opposition of the aqueous and igneous agencies will be more obvious when we come to consider their operation in the construction of the earth in the ages long gone by.

**194. Volcanoes.**—The most striking present manifestations of the agency of heat we have in volcanoes. These are the grand chimneys of the earth; and, as vents for the great furnace of fire in its interior, undoubtedly save the earth's crust from most disastrous consequences. The manner in which the volcanic mountains were constructed will be shown in the chapter on the formation of the earth. Volcanoes are *extinct* or *active*, the extinct being those mountains which, from their shape and composition, are known to have been once in operation, though not since the advent of man, and the active being those which have been eruptive since his advent. While some active volcanoes are continually active, most of them have seasons of eruption, with intervals of rest of various lengths. The shape of a volcano is more or less conical, the cone being truncated, that is, without a top. When it is inactive, the cavity of the crater or opening is shut up with a crust of solid lava. It is supposed that the occasion of an eruption is the introduction of water in some way into the interior, generating steam. The expansions of the steam produce the earthquakes which so commonly precede an eruption. The steam at length accumulates to such an extent that it bursts the solid cover of the crater, scattering its fragments and dust aloft, with sheets of flame, followed by the overflow of the lava. The steam which escapes forms above the volcano a bright cloud, and with it there are continual discharges of lightning, with thunder, the explanation of which has been found in the fact that steam escaping from a boiler has decidedly electrical properties. The dust and the condensed steam produce, of course, showers of rain and mud in the neighborhood. The matters thrown out from a volcano are various in their character. When the eruption is about to terminate showers of cinders fall, and the last of the eruption is a mixture of smoke and vapor.

**195. Vesuvius.**—There is no record of the action of

this volcano previous to the Christian era, although it had the structure of a volcanic mountain. It had been so long inactive that vines grew all over the interior of the crater. In A.D. 79, after a series of shocks during a dozen years previous, an eruption occurred which buried up the cities of Herculaneum and Pompeii. There was not much lava thrown out in this eruption, but chiefly loose material, such as sand, ashes, cinders, and stones. The steam which rose at the same time from the volcano, being condensed in the air, fell in showers, and, being mingled with the ashes and sand, currents of mud ran into the streets, houses, and cellars, filling them up. "Hence it is," says Hitchcock, "that when these cities were first excavated, more than a hundred years ago, every thing enveloped was in a most perfect state of preservation—the pavements of lava, with deep ruts worn by the carriage wheels; the names of their owners over the doors of the houses; the frescoed paintings as bright as though put on but yesterday; fabrics in the shops still showing their texture; vessels of fruit so well preserved as to be easily recognized; bread retaining the stamp of the baker, and medicine yet remaining on the apothecary's counter. The whole constitute perfect examples of fossil cities." Some skeletons have been found incased in *tuff* or *tufa* (porous volcanic rock), finely preserved. Around the neck of one was a chain of gold, and on the bones of the fingers were jeweled rings. Resina was built over the buried Herculaneum, but this was destroyed by a river of lava in 1631. In Fig. 44 you have represented the appearance of the crater of Vesuvius in 1829. The top of the volcano was blown off in 1822 to the extent of more than 800 feet.

196. **Etna.**—This volcano, on the island of Sicily, is about 90 miles in circumference, and is nearly two miles high. In its eruption in 1669 its lava destroyed fourteen towns and villages before arriving at Catania, and there, although the wall was 60 feet high, it accumulated to

Fig. 44.

such an extent as to pour over it, and, after destroying a part of it, ran on a distance of 15 miles, and then emptied into the sea. Mantell says of one of Etna's eruptions, "If any person could accurately fancy the effect of 500,000 sky-rockets darting up at once to a height of three or four thousand feet, and then falling back in the shape of red hot-balls, shells, and large rocks of fire, he might have an idea of a single explosion of this burning mountain; but it is doubtful whether any imagination can conceive the effect of one hundred of such explosions in the space of five minutes, or of twelve hundred or more in the course of an hour, as we saw them."

197. **Crater of Kilauea.**—This volcano, which is always active, is on the island of Hawaii. It is the most remarkable and singular volcano in the world. It is not a truncated cone, but a crater situated on high land near the

base of Mount Loa. It is a chasm eight miles in circumference, and situated in the midst of another chasm still greater in extent, that is surrounded by a precipice varying from 200 to 400 feet in height. In the midst of the inner chasm stand up fifty or sixty conical craters, many of which are continually in action. Rivers of melted lava run about among the craters, and red-hot stones, with cinders and ashes, are sent up with flame from those which are active, in some cases to an immense height.

198. **Tomboro.**—This is a volcanic mountain in the island of Sumbawa. Lyell thus speaks of an eruption of it in 1815: "It began on the 5th of April, and was most violent on the 11th and 12th, and did not entirely cease till July. The sound of the explosions was heard in Sumatra, at the distance of 970 geographical miles in a direct line, and at Ternate, in an opposite direction, at the distance of 720 miles. Out of a population of 12,000, only twenty-six individuals survived on the island. Violent whirlwinds carried up men, horses, cattle, and whatever else came within their influence, into the air; tore up the largest trees by their roots, and covered the whole sea with floating timber. Great tracts of land were covered with lava, several streams of which, issuing from the crater of the Tomboro Mountains, reached the sea. So heavy was the fall of ashes that they broke into the president's house at Birna, forty miles east of the volcano, and rendered it, as well as many other dwellings in town, uninhabitable. On the side of Java the ashes were carried to a distance of 300 miles, and 217 toward Celebes, in sufficient quantity to darken the air. The floating cinders to the windward of Sumatra formed, on the 12th of April, a mass two feet thick and several miles in extent, through which ships with difficulty forced their way. The darkness occasioned in the daytime by the ashes in Java was so profound that nothing equal to it was ever witnessed in the darkest night."

199. **Graham's Island.**—This island rose up out of the

sea off the island of Sicily in the year 1831. Fig. 44, *a*, represents it as it appeared on the 18th of July. On the

Fig. 44, *a*.

4th of August it appeared as seen in Fig. 44, *b* (p. 116). It was 180 feet high, and over a mile in circumference. It was composed mostly of loose material, and therefore was so far washed away by the water in the course of two or three years that there was nothing left but a rocky shoal. Many other islands have been seen to rise out of the sea, and there are also many which are composed of lava, showing that they were produced, we know not

Fig. 44, b.

when, by volcanic action. There is a group of such islands in the Grecian Archipelago, the advent of some of which is indeed known. A new island was thrown up near Iceland in 1783, consisting of high cliffs, and from various parts of it there were emitted fire, smoke, and pumice. This island was taken formal possession of by his Danish majesty, but before a year had passed there was nothing left to show where it was but a reef of rocks.

200. **Earthquakes.**—As earthquakes very generally precede an eruption of a volcano, and cease when the lava pours forth, the conclusion is legitimately arrived at that all earthquakes are caused principally by the movements of pent-up volcanic matter. When they occur at a considerable distance from volcanoes, they are owing to heavings of that vast body of melted matter which is contained in the interior of the earth; and it has been observed in such cases that, when there has been a succession of earthquakes, they have ceased when some volcano has, by its eruption, relieved the pressure, or when some earthquake of great violence has had the same effect. Thus, in 1811, there were many earthquakes in South Carolina, which ceased altogether when Caraccas

and Laguyra, in South America, were destroyed. Another cause of earthquakes has been supposed to be a bending in the earth's crust, arising from changes of temperature. On this point Professor Dana gives the following familiar illustrations: "All are familiar with the cracking sounds occurring at intervals in a board floor of a house, and arising from change of temperature, especially in a room in winter that is heated during the day; and with the more common sounds of similar character from the jointed metallic pipe of a stove or furnace, given out after a fire is just made, or during its decline. In each case there is a strain or tension accumulating for a while from contraction or expansion, which relieves itself finally by a movement or slip at some point, though too slight a one to be perceived; and the action and effects are quite analogous to those connected with the lighter kind of earthquakes." Besides the vibration or wave movement produced in the earth in an earthquake, there is also a vastly more rapid vibration, which causes the sensation of sound. The latter vibration commonly extends much farther than the former.

201. **Effects of Earthquakes.**—Earthquakes produce various effects according to their violence, extent, and accompanying circumstances. They are chiefly fractures of the earth's crust, sometimes very extensive; displacements, either elevations or depressions for the most part; the draining of lakes, and the production of new lakes; the production of waves in the sea, sometimes to a great distance; and the destruction of life in fishes from the mere shock that is given to them. The displacements and fractures often involve a great destruction of human life. In the famous earthquake at Lisbon in 1755, the greater part of the city was thrown down, and 60,000 persons were killed. Some most remarkable elevations and depressions have attended earthquakes. The coast of Chili was in 1822 raised three feet over a space of 100,000 square miles, an area equal to half of France. In

the year 1772, while Papandayang, one of the loftiest volcanoes in the island of Java, was in eruption, the mountain, to the extent of 15 miles in length and 6 in width, fell in with all its inhabitants, and wholly disappeared.

202. **Solfataras.**—There are localities in the neighborhood of volcanoes, or where volcanoes formerly existed, from which sulphur vapors arise, and result in incrustations of sulphur. Carbonic acid gas also often escapes from the action of some acid, as the sulphuric, upon carbonate of lime or limestone.

203. **Geysers.**—Hot springs are very apt to occur in volcanic neighborhoods, and when they are found where there are no volcanoes, as in Virginia, there is evidence in the character of the rocks that volcanic action has been present at some time in the past. The famous hot springs of Iceland are called geysers. In the Great Geyser (Fig. 45) we have, in a mound of silicious rock, a basin-shaped cavity about fifty feet in diameter. From this cavity there goes down perpendicularly into the earth a pipe some eight or ten feet in diameter to the depth of seventy-eight feet; and in the eruptions of the geyser there issues from this a huge column of water to the height of 150, sometimes 200 feet. As the water mounts up it is divided into numberless jets, and descends in the form of spray, making an immense and splendid fountain. The basin is sometimes empty, but is usually filled with beautifully transparent water, boiling briskly. When the boiling is violent, and especially when the issuing stream throws up the water, subterranean noises are heard like the booming of distant cannon, and the earth is slightly shaken. Each eruption is terminated by a column of steam, which shoots upward with a thundering noise. The hot water, from causes alluded to in § 116, dissolves some of the silica in the rocks within, and as it falls deposits this, and thus forms the mound. Much of the silicious matter is also deposited in the region round about, the moisture being diffused

Fig. 45.

by the winds blowing upon the descending spray. Wood is often found in the neighborhood petrified from this cause.

204. **Explanation.**—The action of the geyser is intermittent, and its explanation has been thus given by Lyell: "Suppose water percolating from the surface of the earth to penetrate into the subterranean cavity A D (Fig. 46, p. 120) by the fissures F F, while at the same time steam at an extremely high temperature, such as is commonly given out from the vents of lava currents during their solidification, emanates from the fissures C

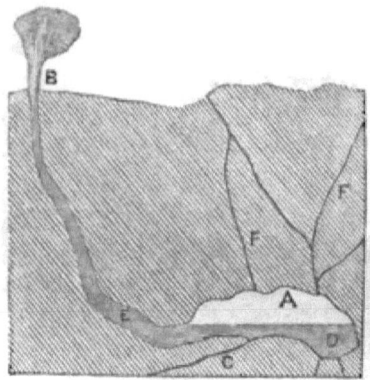
Fig. 46.

C. A portion of the steam is at first condensed into water, while the temperature of the water is elevated by the latent heat thus evolved, till at last the lower part of the cavity is filled with boiling water, and the upper part with steam under high pressure. The expansive force of the steam becomes at length so great that the water is forced up the fissure or pipe E B, and runs over the rim of the basin. When the pressure is thus diminished, the steam in the upper part of the cavity A expands until all the water, D, is driven into the pipe; and when this happens, the steam, being the lighter of the two fluids, rushes up through the water with great velocity. If the pipe be choked up, even for a few minutes, a great increase of heat must take place; for it is prevented from escaping in a latent form in steam, so that the water is made to boil more violently, and this brings on an eruption."

205. **Sand Moved by Wind.** — In some parts of the world great changes are effected by clouds of sand. The sand-hills produced in this way are called *dunes* or *downs*. Very commonly they come from sand that has been washed up upon the shore by the sea, which, on being dried, is carried inland by the winds. There are many dunes of this kind in Cornwall, England, and on the coast of France a great number of villages have been entirely destroyed by them. They are quite common on some parts of the coast of the United States, and especially on Cape Cod. Sometimes dunes occur in the interior of a country. In Egypt the westerly winds have blown the sand over almost the whole of the country west of the Nile, making it a desert, and the remains of

ancient cities, with their temples and palaces, are found covered in the sand.

**206. Alteration of Levels.** — Besides the depressions and elevations which, as you have seen, have been produced violently and suddenly by earthquakes, there are other changes of a similar character which have been very gradual. There is ample proof that in Sweden there have been extensive alternate elevations and depressions, which have gone on so slowly that no disturbances have resulted. One of the most interesting evidences of change of level is found in the pillars of what has been supposed to be the Temple of Jupiter Serapis, near Naples. They are represented in Fig. 47. You

Fig. 47.

see some way up the pillars a roughness of the surface. This results from a collection of just such perforations as are known to be made by certain sea mollusks called
F

lithodomes, the name coming from two Greek words, *lithos*, stone, and *domos*, house, as these animals dwell in the holes which they make in the stone. This shows that the pillars were once, for a considerable time, immersed in the sea as far as the upper limit of the roughness, which is 23 feet. Here, then, was a depression of the land, after the temple was built, to a depth of over 23 feet, and it was effected so gradually that the pillars were left quietly standing. The subsequent elevation must have occurred with the same slowness also. The fact that Scotland has been raised up from 15 to 30 feet since man came upon the scene is very beautifully demonstrated by Hugh Miller in his Lectures on Geology. The old coast-lines have been traced by him lying above the present coast-lines, the identity between them, he remarks, being as decided as that "between two contiguous steps of a stair, covered, the one by a patch of brown and the other by a patch of green, in the pattern of the stair-carpet." One of the most extensive and remarkable of the gradual changes of level that have taken place has been observed in Sweden and Norway. The evidences of it having been long noticed in a rude way by fishermen and pilots, men of science at length undertook the investigation of it systematically, by marking the rocks in different places, and observing them from time to time. In this way it has been ascertained that there is a rising of the land in the northern part of this region, and a sinking of it in the southern part, and the rate is such at some places that the movement would amount to several feet in a century. And that this movement has been going on for many centuries is shown by the fact that such shells as are now common in the Baltic Sea are found inland, beginning at the shore all along up to even 400 feet above the level of the sea. Other similar changes have been noticed in South America, in Greenland, and in other countries.

207. **Organic Agencies.**—Rocks are continually form-

ing to a large extent from material provided by organic agencies—that is, by the agency of animals and vegetables, especially the former. For example, the carbonate of lime that is dissolved in water is taken up by animals —coral animals, shell-fish, etc., and, forming their framework—their skeleton, as we may term it—is afterward deposited as mineral matter, becoming again, as it was originally, a part of the solid substance of the crust of the earth.

208. **Alterations made by Man.**—Wherever man fixes his habitation he effects more or less of change in the earth's surface. These changes are especially manifest in cities where large bodies of men are congregated. Here levels are often considerably altered, as may be learned from the recollections of the oldest inhabitants. But all the changes produced by man are almost as nothing compared with those which come from the operations of the natural causes that I have noticed in this chapter.

---

## CHAPTER XI.

### CONSTRUCTION OF THE EARTH.

209. **Stages in the Construction of the Earth.**—You will see, in the developments which I shall make in this and the succeeding chapters, that there has been a succession of changes, each one occupying a long period of time, the earth being brought gradually into a proper condition as a habitation for man. You will see that the continents were once mere germs of continents, and that they grew to be what they are after a manifest plan, the steps of which the geologist has been able to some extent to discover by his researches among the rocks. You will see that during all this time there was much of tearing down and rebuilding as a part of the plan, the sedi-

mentary ruins of rocks being the materials out of which a large portion of the rocks we now find were constructed. You will see that, although at the first there were ages in which there was no life, vegetable or animal, there was a long series of ages before man appeared in which the earth swarmed with life, the relics being found now imbedded in the rocks, differing from each other, however, in a marked manner, as geologists have found, in the different stages of the earth's construction. It is from these differences in the forms of life that geologists have been able to mark out the periods or ages of the world's formation. In the present chapter it is my intention to point out some of the processes by which the earth has been gradually built up into its present condition, as preparatory to the consideration of what took place in each of its several ages. What you have already learned of the present changes going on in the earth, you will find, will throw much light upon this subject, because, as I have before stated, the same agents which are at work now in these changes did the work in the changes of the far past.

210. **Nature of Geological Evidence.**—The geologist takes the results of processes which are now going on in the earth, and, comparing them with the results which he finds buried up in the rocks, adopts his conclusions in regard to the latter, and does so without danger of error if he be properly cautious. I will give a few illustrations of his modes of reasoning.

The geologist finds a rock which, on examination, is discovered to be of the same composition with clay, and, more than this, has similar layers. He infers that the rock was once clay, and in some way became solidified or changed into rock. By the same reasoning he infers that certain rocks were once mud, and certain others were once sand. The inference in these cases is confirmed by the fact that the solidification has sometimes been known to take place within a short period of time, as you have already seen in § 180, and will see more fully hereafter.

The geologist finds a rock which contains imbedded in it certain shells. These shells were, like the shells we now pick up on the sea-shore, once inhabited by animals. But how did they get into the rock? This question is easily solved by the geologist. He finds shells at the present time imbedded in mud or sand, the mineral character of which is precisely the same with that of the solid rock containing the other shells. He then justly infers that the latter were once in similar mud or sand, which is now solidified.

He makes a farther inference in regard to this rock by comparison with some other. Some of its shells may be of the same species with some that are found at the present day, while those found in the rock with which it is compared may all be different from any of the present species. His inference is that the latter rock belongs to an earlier age than the former. It may have been made or solidified into rock hundreds, or even thousands of centuries before the other, though the two may now lie in juxtaposition. The kind of observation here indicated, you will find as we proceed, is largely made use of in determining the relative ages of rocks; for as, during almost all the ages of the earth, there have been living beings, but differing in character from age to age, the remains of life found in the rocks differ according to the ages in which the rocks were formed.

Take another case. You often see in a block of sandstone—a step perhaps at some door—pebbles imbedded in the material of which most of the rock was made. These pebbles are such as you have seen on a shore, and you know that they once were rough pieces of rock, and that they were smoothed by being rubbed together a long time, as those you see on the shore have been, by water rushing over them, and that after this was done they became mingled with sand, and the whole became a solid rock. This pudding-stone, as it is called, we can make such inferences about just as clearly as we can infer the mode of making a plum-pudding from its appearance.

The geologist goes much farther than this. He discovers often in the layers of rock tracks, and even the marks of rain-drops and ripples, made perhaps ages upon ages ago. I will give a single example. Professor Dana says of some slabs examined in Pottsville, Pennsylvania, "We thus learn that there existed in the region about Pottsville at that time (a period just before the coal of that region was deposited) a mud flat on the border of a body of water; that the flat had been swept by wavelets, leaving ripple marks; that the ripples were still fresh when a large amphibian walked across the place; that a brief shower of rain followed, dotting with its drops the half-dried mud; that the waters again flowed over the flat, making new deposits of detritus,* and so buried the records."

211. **Classes of Rocks.**—First, rocks are divided into two grand classes, the stratified and the unstratified. The stratified appear in strata or layers, and the surfaces of these strata are nearly or quite parallel. They are either earthy aggregates, as sandstones, or simple chemical precipitates from solutions, as the limestones sometimes are. Those which are mere aggregates were deposited as sediment, and therefore are called sedimentary rocks. They are also called *aqueous* rocks, as are also those which were precipitated, because water was the agent by which the matter composing them was brought to the locality and deposited. You have an example of stratified rock in Fig. 48. Stratified rocks very generally contain fossils—that is, remains of plants and animals which were in existence at the time that the material of which they are composed was deposited, and are there-

---

* This term is applied to what has been removed from the surfaces of rocks by the erosion of water and other causes. When the material thus removed is coarse, it is called *débris*.

Fig. 48.

fore called *fossiliferous* rocks. *Metamorphic* rocks are stratified rocks which have been changed by the action of heat, and perhaps some other auxiliary agencies. In this alteration, any fossils that the rocks originally contained are obliterated, the material which composed them having been altered with the rest of the rock in the arrangement of its particles. Thus limestone containing shells and corals has been often converted into granular limestone or marble, a crystalline texture being thus given to the whole. The term metamorphism means transformation, and comes from two Greek words, *meta*, which is the same as *trans* in the Latin, and *morphe*, form. A very decisive proof that heat is the principal cause of metamorphism we have in the fact that great artificial heat, if long continued, changes the structure of stones. Sandstone used in furnaces has sometimes been known to become crystalline—that is, has been metamorphosed.

The *unstratified* rocks are not divided into parallel layers, but they are commonly a shapeless mass, as seen in Fig. 49 (p. 128). They were not deposited from sediment, but were formed under the influence of heat, and were thrust up from within the crust of the earth. They, of course, never have any fossils in them. They general-

Fig. 49.

ly appear in mountains, forming often their central part, while the stratified abound in plains, or flank the sides of the mountains. When a mass of unstratified rock, as granite, stands up as the axis or central part of a mountain, stratified rocks commonly slope off from the granite, this having thrust them to the one side and the other as it rose up out of the bowels of the earth.

There is one class of unstratified rocks, the trappean, noticed in § 152, which have a tendency to regularity of shape, and in some cases the tendency is fully carried out. In this they are distinguished from the shapeless masses of granite and other unstratified rocks.

212. **Stratification, Lamination, Joints, and Cleavage.**— The word *layer* is often used as meaning the same thing as *stratum*, but in the strict use of the latter term it includes all the layers of the same kind which are next to each other. A *stratification* means a succession of layers, either of the same kind or of different kinds. Layers differ very much in thickness, according to the manner in which they were laid down. When they are very thin they are called *laminæ*. In shales and micaceous sandstones the laminæ are so thin that we may properly speak of them as films, and it is plain that in such cases the stratum or bed was formed by the very gradual accumulation of films of clay or of micaceous spangles, which settled down at the bottom of comparatively still

water. Each film or lamina was the result, of course, of a separate deposition—that is, after each lamina was completed, there was a pause for a time in the deposition from the water. In those cases in which there are fifty, or even a hundred laminæ in the thickness of an inch, the process, including the intervals, must have been very slow. Some years must have been required to lay down a foot of such a bed. Where the lamination is not so fine, the accumulation of the sediment of which the rock was made was more rapid. The term *formation* is applied to a series of strata that have a relation to each other in similarity of fossils, and which are therefore included in the same age or period. This is the general idea, but the word is used rather loosely, being made to refer to a larger range of strata at one time than at another. In looking at a stratified rock, you observe beside the horizontal lines which indicate the divisions between the layers certain lines which cross these. The two sets of lines are seen in Fig. 50. The lines which

Fig. 50.

run up and down across the layers indicate planes of division, which are called *joints*. Some of these, you see, extend farther than others, and these are called *master-joints*. There are commonly two sets of joints at right angles to each other, as is represented in the figure. It

F 2

is this arrangement that enables the workman to get out readily square, or, rather, squarish blocks of stone. The regularity of these divisions varies much in different cases. Sometimes it is so great that the upper surface of a stratum has very much the appearance of a regular pavement made with nicely-fitted slabs. Where, from any cause, the rocks are occasionally thrown down, exposing new surfaces of the strata with their joints, there is a resemblance to the ruins of fortifications. In Fig. 51 we have a view of some of the cliffs of Cayuga Lake,

Fig. 51.

N. Y., which weathering and the undermining action of the water are continually wearing away, so that fresh surfaces show in very definite manner the planes and lines of division. *Cleavage*, or the slaty structure, is another kind of division in rocks. This is sometimes parallel with the layers, and sometimes runs across them. Both the larger divisions by joints, and the smaller ones by cleavage, each often extend over vast regions of country at the same angle, showing that some causes operating extensively produce them.

213. **Order of Succession of Rocks.**—There is a regular order in the stratified rocks which is never trans-

gressed, so that the geologist can apply conclusions made in regard to rocks in one quarter of the world to those which exist in any other quarter. There may be omissions, but there is never any change in the order. "As a bookbinder," says Phillips, "sometimes neglects to bind in a particular leaf, so Nature sometimes omits a particular rock; but she never misplaces the rocks, as the careless workman sometimes misplaces the pages." There is a very signal example of such an omission in the rocks of this country. We have here the cretaceous or chalk formation—that is, the series of rocks so called; but the chalk, which forms so prominent a part of this series in England, France, and many other countries, is wholly absent; that leaf in the American geological volume is left out.

214. **Flexures of Strata.**—Though strata are deposited generally either horizontally or nearly so, they are often very much bent, as is represented in Fig. 52. There are

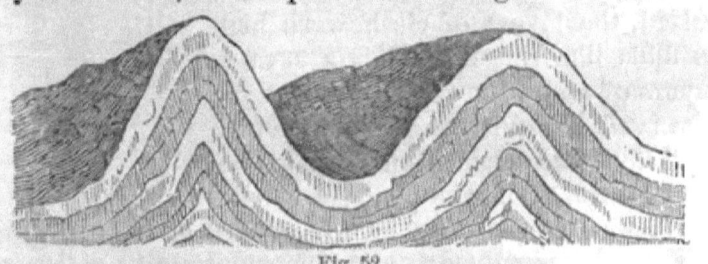

Fig. 52.

some magnificent exhibitions of these flexures in some of the mountainous regions. Among the Alps there are mountains thousands of feet high, looking, as Professor Hitchcock says, "as if crumpled together by some mighty hand." It would seem as if it were not possible thus to bend rocky strata without breaking them. But probably the rocks were both moistened and heated when this was done. Besides, the bending force undoubtedly was made to act very gradually. Some light is thrown on this point by some observations in regard to the bending of ice. Though ice is a very brittle solid, large blocks of it have been known gradually to bend very considerably. Kane saw this in a flat block of ice eight feet thick, and twenty or more wide, supported only at its two ends. In the course of two months it became so much bent that the middle of it was depressed five feet.

215. **How the Flexures are Produced.**—It is supposed that the flexures of the strata were produced by lateral pressure. This has been illustrated by a simple experiment by Sir James Hall. He took pieces of cloth, some linen and some woolen, and, placing them in layers on a table (Fig. 53), compressed them by the weight $a$. He then applied pressure to the sides $b\,b$, as shown in Fig. 54, and found that while the weight $a$ was raised, the layers of cloth were bent in folds like those which are seen in layers of rock in nature. Sir C. Lyell observed a result near Boston which well illustrates the action of this lateral pressure. For the purpose of converting part of an estuary, overflowed at high tide, into dry land, a vast quantity of stones and sand was thrown into it. The effect of this was to push up the bottom of the estuary alongside of this load of stones and sand, so that it, in the course of a few months, was raised six feet above high-water mark, and some five or six folds were produced in it, just like the folds, small and great, which the geologist finds in rocky strata.

Fig. 53.

Fig. 54.

216. **Upheavals of Strata, with Fracture.**—Fig. 55 is a

Fig. 55.

CONSTRUCTION OF THE EARTH. 133

representation of such an upheaval. Fissures of various dimensions, sometimes of immense size, are made in this way, and this is the origin of many of the valleys which we see with steep, bold sides. Chasms and caverns are produced by such upheavals; water, by its erosive agency, acting, however, at the same time, and afterward in many cases, especially where there are caverns of considerable extent. Natural bridges are the result of upheavals, an example of which is given in Fig. 56, representing that of Icononzo, in South America.

Fig. 56.

217. **Strata at different Angles.**—One consequence of upheavals with fracture is that we have strata lying at different angles, as is often shown when their projecting ends or sides are laid bare. The order of the rocks is thus brought to view sometimes to a great extent, as in the case of Snake Mountain, in Vermont, represented in Fig. 57 (p. 134). The place of fracture is at *c*, and there is brown clay at *e*, extending from here west six miles to Lake Champlain, and lying upon *n*, limestone rock. I will trace the layers of rock up the mountain. At *b* we

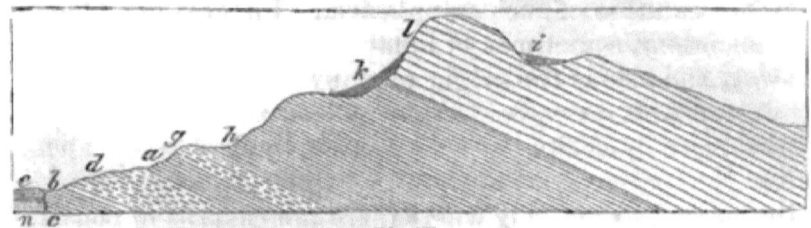

Fig. 57.

have sand rock, at *d* one kind of limestone, at *a* another kind, and the same as that on which the brown clay, *e*, lies; at *g* is slate rock, at *h* are shales, and at *l* red sand rock; at *k* is débris from the sand rock, *l*, and at *i* is a cranberry meadow lying on peat. Sometimes the strata in the upheaval become vertical, as seen in Fig 58, which

Fig. 58.

is a representation of vertical strata on which a castle has been built. These strata were laid down horizontally far back in the past, ages upon ages before man was introduced, and then by some mighty upheaval were raised to their present position. Sometimes strata appear to be horizontal when they are very far from it. Take such a case as that pre-

Fig. 59.

sented in Fig. 59. If the spectator were in the vessel, so as to look at the rocks as they appear on the side *p*, the strata would appear to be horizontal, when in reality they lie at an angle of about 45°, and gradually become vertical as you go toward *a*.

218. **Dip and Strike.**—These are terms which are much used in geology, and therefore require explanation. The line of variation of strata from a horizontal line is called their *dip*. For example, in the case shown in Fig. 60, if

Fig. 60.

the angle made by the lines of the strata, *b b*, with the horizontal line, *a*, be 45 degrees, these strata are said to dip 45 degrees. The *strike*, on the other hand, is the line of direction in which the edges of the strata run. The meaning of these terms can be made clear by a simple illustration.

Fig. 61.

If you place a book on the table, as seen in Fig. 61, with the edges of the leaves downward, and move one side of the cover away from the body of the book, a line from the back of the book straight down the cover, *b*, will be the line of dip, while a line along the back of the book, *a a*, will be the line of strike.

219. **Anticlinal and Synclinal Lines.**—An anticlinal line is a line along which the strata dip in opposite directions. This may be exemplified by placing the book

Fig. 62.

in the position seen in Fig. 62. The line along the back of the book represents the anticlinal line. The synclinal line is the reverse of this. To represent it, let the book be placed in the position seen in Fig. 63 (p. 136). The line at the angle of the two parts of the book is the synclinal line. These lines do not, however, run along sharp

Fig. 63.

angles, as these illustrations would indicate, but curves, and therefore anticlinal and synclinal curves are spoken of—in the anticlinal the sides inclining upward toward each other like the roof of a house, but in a curved line, while the sides of a synclinal curve incline downward toward each other, like the sides of a trough.

220. **Measuring Strata.**—Where strata outcrop we can very readily take the measurement of their thickness. The manner in which this is done is so well pointed out by Mr. Jukes, Director of the Geological Survey of Ireland, that I will give you his description of it entire: "Having procured a good map of the district we are going to examine, a pocket compass, and a small instrument called a clinometer, by which we can determine the angle at which a bed inclines from the horizontal plane, we begin to look for exposures of rock. The district may perhaps at first appear to be entirely covered by soil and vegetation, but, when thoroughly examined, it may show here and there some crags of bare rock, some bare cliffs on the side of a river, some cutting on a road-side, or some quarries. Let Fig. 64 be a piece of our map traversed by a brook and a road, and suppose that we find some quarries of limestone in the northeast corner of the district, all the beds of which dip to the southwest at an angle of 20°. Let us represent these limestones by the cross-barred lines in the northeast corner of the figure, drawing them in a northwest and southeast direction to represent their strike, and indicating their dip by an arrow with 20° annexed to it. In examining the banks of the brook, suppose we find, in two places lying in the same line of strike, some shales, represented by plain close lines; a bed of coal, represented by a thick dark line; and some sandstones, represented by dots; and, for simplicity's sake, let us suppose them all to dip south-

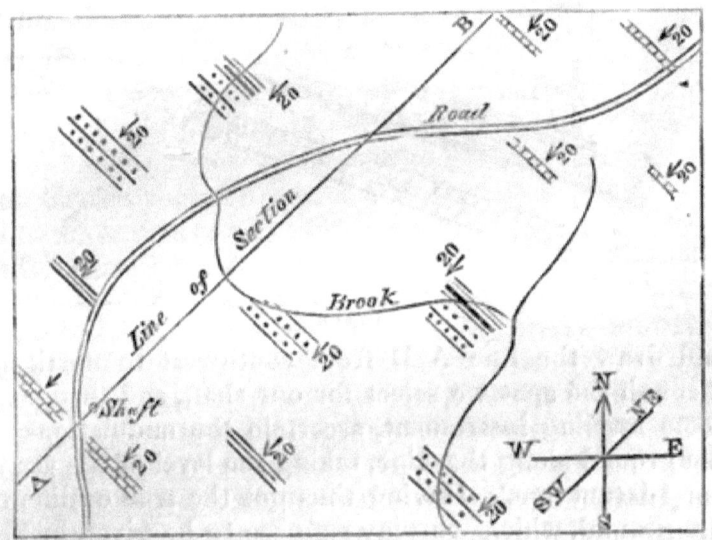

Fig. 64.

west at 20°. Then suppose that we find one or two other detached exposures of sandstone and shale to the southwest of the above, with similar strikes and dips. And, lastly, suppose that in the southwest corner of the district we find some more limestone of quite a different kind from that in the northeast corner, but still dipping to the southwest at 20°. It will be obvious that these limestones in the southwest corner of the map lie above all the other rocks, while the limestones in the northeast corner rise out from beneath all the rest."

221. **Calculating the Thickness of Strata.**—In the case just detailed, we have data for arriving at the thickness of this whole set of strata. The manner in which this is done can be explained on Fig. 65 (p. 138), which represents a section of the district at right angles to the strike of the rocks. "Suppose," says Mr. Jukes, "that we wish to sink a pit at the part marked *shaft* in the map, in search of the coal, and that, before commencing, we wish to know how deep the coal is there, and how far it is to the limestone below it. We take the map,

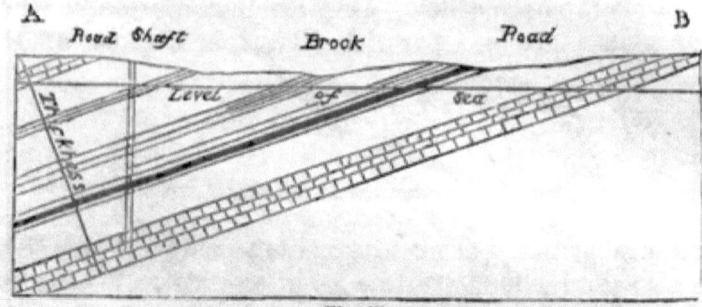

Fig. 65.

and draw the line A B from southwest to northeast through the spot we select for our shaft, and then, with some leveling instrument, ascertain the undulations of the ground along that line, taking the level of the sea as our 'datum line.' Having thus got the true outline of the ground, which we may suppose to be given in Fig. 65, we draw lines inclined at an angle of 20° toward the southwest at the several spots where the section line is cut by the outcrops of the beds, or by straight lines joining them, and this section will then show us the depth at which the shaft will reach any of the beds. Similarly, if we wish to know the thickness of the whole succession of beds, from the highest to the lowest, of those exposed in the district, the length of the line marked *thickness* will give it to us, when measured on the scale we adopt for our map and section."

In these simple operations is indicated the mode of ascertaining some great geological facts, for it is by such observations and measurements that we gain a far more extensive knowledge of the crust of the earth than we can by the deepest mining excavations. The great upheavals have turned up to us the leaves of geological history for our reading, many of which we could not otherwise have read, for they would have been so deep that no exploration of man could have reached them.

222. **Conformable and Unconformable Strata.**—Strata are said to be conformable when their surfaces or planes

are parallel to each other. They are unconformable when their planes are not parallel. Thus, in Fig. 66, all the

Fig. 66.

horizontal strata *a* are conformable to each other, and so are also the inclined strata *b*; but the strata *b* have the strata *a* lying unconformably upon them. All the strata *b* were deposited in a horizontal position, and there was no upheaval till the last of them was completed. Then the state of quiescence was broken, and the strata were lifted up as you see; and when the upheaving force ceased to act, the upper horizontal strata began to be deposited. Periods of immense length were required for all these successive processes, for the upheaval was probably very slow, as the depositions certainly were.

223. **Faults.**—Whenever in a fracture of strata there occurs a dislocation of them, leaving those on one side of the fissure higher up than those on the other, it is called a fault, a name which was first given to such a

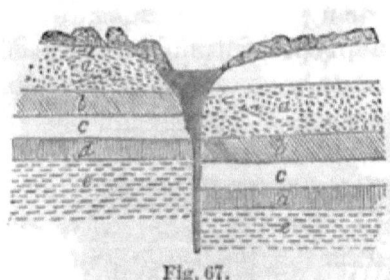

Fig. 67.

state of things by miners. In Fig. 67 you have a representation of a fault. The strata marked with corresponding letters on each side of the fissure were continuous till they were all deposited, and then, by some force, either the strata on the right hand were depressed, or those on the left were elevated. Faults vary much in the degree of dislocation, from a few feet up to many thousands. Some of the grand changes in the earth's crust have been produced by faults.

224. **Denudation.**—This term is applied to the wear-

ing away of rocks by flowing water. This erosive power of water, spoken of in § 186, has produced some of the most immense changes that have taken place in the earth. You have seen what vast quantities of material it supplies to be carried down by the rivers into lakes and seas. The hills and mountains, though made by foldings of strata and upheavals, generally received their final shaping from denudation. Though erosion by water is comparatively a slow process, yet, acting through long periods, it produces great effects. It is supposed that the Appalachian Mountains have lost from this cause as much material as is now contained in them, or even more, and the same can be said of some other ranges of mountains. Indeed, in the changes of the long ages occupied by the earth's foundation, the great majority of the rocks have been built up from materials furnished by denudation from pre-existing rocks, and often denudation and reconstruction have succeeded each other many times over in the case of the same material. This same succession is seen to a considerable extent even now.

225. **Faults with Denudation.**—Denudation has often united with faults in producing great changes. This may be illustrated by Fig. 68, representing a section of

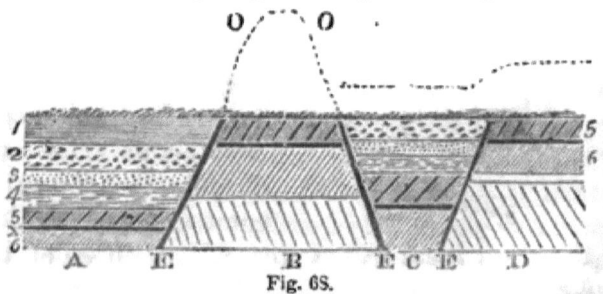
Fig. 68.

a coal-field in England of several miles in extent. There are four divisions, A, B, C, and D, made by faults which have put the strata in them at different relative depths. Thus, in A, the bed of coal, X, is 900 feet from the surface, and has over it various strata of sandstone and shale, 1, 2, 3, 4, and 5. In the divisions B and D there is only the stratum 5 above the coal, which is only 200 feet from the surface, while in the division C the coal lies 700 feet deep. Now, notwithstanding this displacement of the strata, there is no appearance of the upheavals on the surface of the coal-field, but that is level. If the material all remained on the spot, we should have an uneven surface, and at O O there would be an elevation of 700 feet, for that is the amount of uplift in the strata at that part of the field. The inference is that all this great mass of material, indicated by the dotted line, was removed by denudation. Near Chambersburg, Pa., there is a fault 20 miles in length, and the depth of the dislocation is 20,000 feet, and yet a man can stand with one foot on one side of this fracture and the other foot on the other side. What has become, then, of this immense mass of material 20,000 feet in height? "It must have been swept," says Mr. Lesley, who gives the account, "into the Atlantic by the denuding flood." If this had not been done there would have stood there a bold precipice nearly four miles in height and twenty miles in length. Long ages must have been required for water to effect such a denudation. The proximity in which such a fault places rocks of epochs far distant from each other bring to the mind grand and overwhelming conceptions of the vast periods of time which have elapsed in the building up of the world. On the one side of that crevice you step on limestone that was made long ages before the slaty rock that you step on upon the other, and by accident there are lodged in that crevice fragments of a rock of still another epoch between the epochs of those on either side.

226. **How Mountains were Made.**—You are now prepared to see how mountains were built up. They were not all constructed and shaped in the same way, but there is much variety. I will indicate some of the prin-

cipal modes. I have already mentioned the agency of flexure or plication, as illustrated by Fig. 52. Denudation has much to do with the formation of such mountains, as seen in Fig. 69. Here you see on each side

Fig. 69.

strata which, if continued on over the mountain, would very much increase its height; but they were removed probably by the water sweeping over the mountain all the time that it was slowly rising. This is a good representation of the manner in which most of the mountain chains are made and shaped.

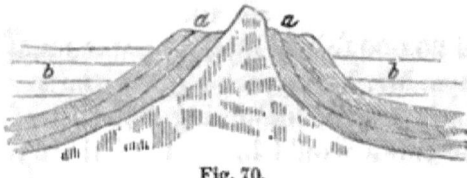

Fig. 70.

Take another example, represented in Fig. 70. Here it is obvious that the strata $a$ were first deposited horizontally; then the middle part or axis of the mountain was thrust up, tilting the strata $a$, and after the tilting was completed the strata $b$ were deposited. In Fig. 71 we

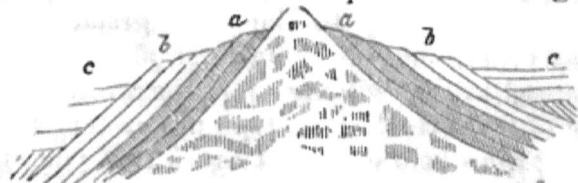

Fig. 71.

see a somewhat different state of things. The strata $a$ and $b$ were both deposited before the axis of the mountain rose, and therefore were tilted up together. Afterward the strata $c$ were formed. Now if the strata $a$ and $b$ are the same in character in the two cases, the mountain represented by Fig. 71 is of much later date in the world's growth than that represented in Fig. 70.

CONSTRUCTION OF THE EARTH.    143

227. **Valleys.**—Intimately connected with the subject of the formation of mountains is that of the formation of valleys. These are chiefly of three kinds: 1. Valleys of *undulation*. These occur between the hills and mountains that are raised by flexures of the earth's strata, as represented in Fig. 52. 2. Valleys of *dislocation*. These are caused by fissures in strata, as represented by Fig. 55. Where the strata are very thick, such valleys are colossal in size and bold in their features. 3. Valleys of *denudation*. These vary much, according as the rocks acted upon by the water are soft or firm. In Fig. 72 is represented an example of valleys formed by de-

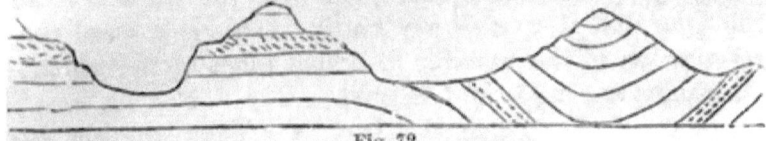

Fig. 72.

nudation. The strata here are not all continuous, but portions of some of them have been worn away by water, leaving the hills with valleys between them.

228. **Volcanoes.**—These are constructed in a different way from common mountains. They are made out of the matters which are ejected from them, so that, as long as a volcano is active, the work of construction is to some extent going on. The solid matters ejected are chiefly lava, cinders, and cinders moistened and forming in their solidification what is called *tufa*. The tendency in the construction is to the form of a cone, which is sometimes very perfectly carried out, as seen in the grand volcano of Japan, Fusiyama, the summit of which is represented in Fig. 73 (p. 144). It stands 14,000 feet high. It has in its top an oval opening, one diameter of which is 3300 feet, and another 1800 feet, and the depth is 1000 feet. This volcano has long been inactive, the latest eruption recorded having occurred in 1707. But two months in the year, July and August, is the summit sufficiently free from snow to permit the ascent, and then the natives are

Fig. 73.

very busy in making their pilgrimages to the mountain, which they consider very sacred, and call "matchless." The steepness of the sides of volcanoes depends on the material of which they are formed, being steepest when the material is cinders, and least so when it is lava. Most frequently volcanic cones are of a mixed character, having layers of lava and of consolidated cinders mingled together. This is represented in Fig. 74, which is intended

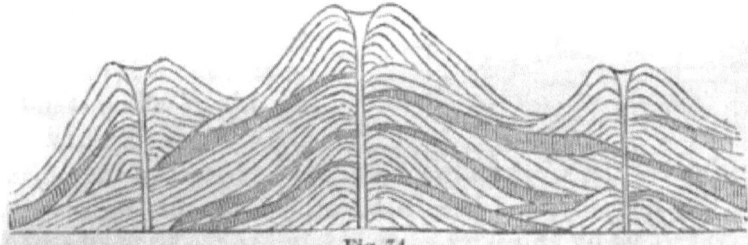

Fig. 74.

to be a section through some volcanic cones. The parts made of short vertical lines represent the layers of lava, and the long lines represent the layers of ashes and cinders. Irregularities occur in the forms of volcanic cones from various circumstances. For example, a wind prevailing in one direction may make ejected cinders accumulate on one side of a crater more than on another. The edge of a crater may be thrown over, as in the case of Vesuvius (§ 195). Lava may burst out through a fissure, and, solidifying, cause accumulation on one side of the cone.

229. **Trap Rocks.**—Though these rocks have a decided resemblance to volcanic rocks in their composition, the eminences formed by them, some of which are lofty enough to be called mountains, were constructed in a very different manner. They are igneous, and yet not volcanic. They were thrust up from below, a molten mass pushing up the superincumbent strata before them, and gradually became solid by cooling. They came up through fissures in rocks, forming dikes, so called, in these fissures. If the trap was not so hard as the rocks

in which the fissure was made, then the denudation

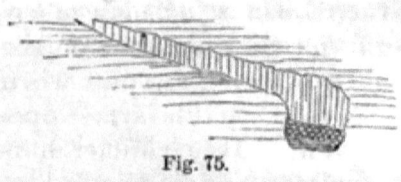

Fig. 75.

which followed left what is called a sunk dike, as seen in Fig. 75. But if, as is usually the case, the fissured rocks were softer than the trap, it left a raised dike, as represented in Fig. 76. Hugh Miller, in

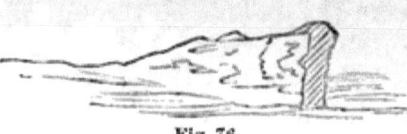

Fig. 76.

commenting upon the trap eminences or dikes in and about Edinburg, on one of which stands the Castle of Edinburg, compares the work of the denuding agencies to the work of the sculptor, because, as he brings out his figures in *alto relievo* by chipping away around them, so have these agencies brought out in bold relief the grand trap prominences by scooping away the soft shales and sandstones which flanked them. The same can be said of the twin prominences, East and West Rock, that add so much to the scenery about New Haven. Sandstone was all about them, perhaps even covered them, until the sculptor, water, with its tides, and waves, and currents, removed it. "Trap scenery," remarks Hugh Miller, "may be described generally as eminently picturesque. From the circumstance that its eruptive masses rise often from amid level fields, and that its hard, abrupt beds, dikes, and columns alternate often with rich, soft strata, that decompose into fertile soils, it abounds in striking contrasts. The soft plain ascends often at one stride into a hill fantastically rugged and abrupt, and bare and fractured precipices overtop terraced slopes or level platforms rich in verdure."

230. **How the Trap Rocks were Formed.**—These rocks were not formed below and then lifted into their places, but, as I have before stated, the material in a molten state was thrust up, and there cooled off and solidified

under the pressure of rocky strata, which it had lifted up, and which was covered with water probably of considerable depth. This could not happen without creating great commotion, especially if there were fractures in the superincumbent strata, letting down the water upon the surface of the red molten mass. Hugh Miller, in his magnificent description of what may be imagined to have occurred in the emergence of the trap rocks, speaks of steam and flame issuing from the fissures up through the boiling waters. This may all be correct; but when he speaks also of the heavens being dark with ashes, as if a real volcano were in action, the facts which the rocks now reveal to us hardly warrant such imaginings. Though fire is the agent in both cases, the formation of trap rocks differs very decidedly from the building up of a volcanic cone. The open throat or chimney, through which materials of various kinds are thrown up from the volcano, is wanting when the trap rocks are elevated into their position.

It is in the cooling and solidification of the molten mass that the prismatic, pillar-like form is sometimes so perfectly assumed, as exhibited in the Giant's Causeway, Fingal's Cave, etc. The igneous origin of these prisms fairly entitle them to the name of "furnaced pillars," which the Ettrick Shepherd has given them.

231. **Veins.**—There are found in the rocks what are called veins—that is, fissures filled with rocky material, or with metallic ores, or with both. When they are filled with rocky material, this is sometimes the same with the rock itself in which the veins are, and sometimes it is of a different composition. Sometimes one vein traverses another, as is shown in Fig. 77. This is often the case with veins of granite. Such

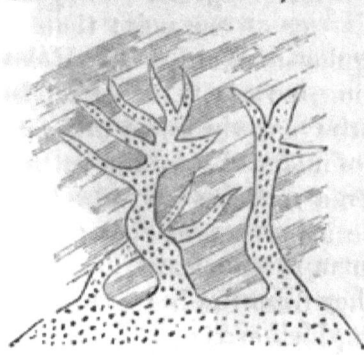

Fig. 77.

veins have often penetrated through the overlying deposits, and flowed over the rocks which they have displaced, as shown in Fig. 78. Sometimes one vein is displaced or faulted by another, as represented in Fig. 79. Here is an

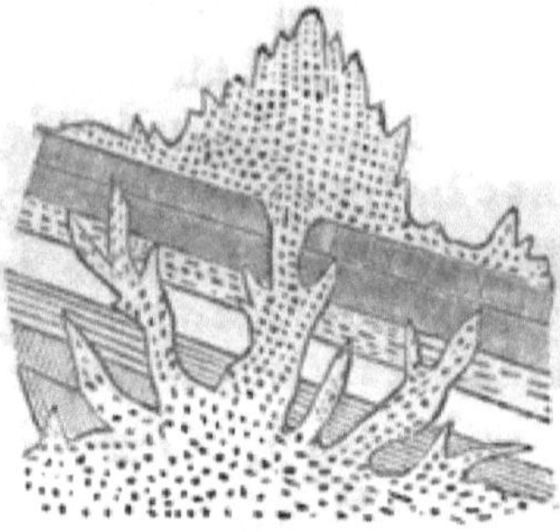

Fig. 78. 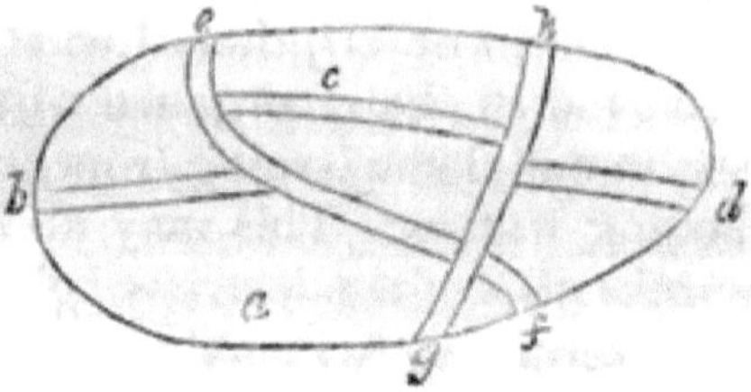    Fig. 79.

outline of the section of a boulder found by Prof. Hitchcock in West Hampton, Mass. It is a fragment of a granite rock which was traversed by three granitic veins. These, as seen in the boulder, are *bcd*, *ef*, and *gh*. The vein *bcd* was made first. The other two were made afterward, displacing or faulting the first, as you see in the figure. As the three veins were made at three different times, they are three different varieties of granite. Here, then, is a record of four different formations of granite, the rock itself and the three veins, and how long the intervals were between them we know not—they may have been ages.

Veins differ from dikes in various respects. They are apt to be irregular in shape, larger at one point than at another, while dikes are regular and uniform. Veins are often compound, containing a variety of materials, sometimes a considerable variety, while dikes are simple, containing only one kind of material, as some volcanic mineral or trap. And when a vein has only one kind of material, this never has the arrangement that a dike often presents—viz., a columnar arrangement, as if one block were placed above another from the bottom to the top of the dike. Those veins which are compound have a very different arrangement from the columnar one of the dikes. It is a banded arrangement—that is, the materials are in bands parallel with the walls of the vein. The bands may be few in number or numerous. If there be metallic ore in a banded vein the rocky material is called the *vein-stone* or *gangue*.

Dikes are due to the injection upward of melted material into openings or fissures in the rocks; but the manner in which veins have been produced is by no means as yet fully understood, and there have been much discussion and speculation about it.

The distinctions between dikes and veins are generally recognized among geologists as I have pointed them out, but sometimes there is a little confusion in the use of these terms. The term *lode* is often given to veins, but seldom to any except those which are metallic.

232. **Drift.**—This term is applied to fragments of rocks which have been scattered over some portions of the earth's surface by other means than the flowing of water. The fragments vary in size from grains up to those which are many tons in weight. When they are of any considerable size they are called boulders. They are supposed to have been scattered by means of icebergs or glaciers, or both. Drift is not found at all in tropical climates. In North America it appears in Canada, New England, the State of New York, and in all the states west of that region in the same latitude. In Europe also it is confined to the northern part. From comparison of the drift with the stationary rocks, and from other observations, it is manifest that all the drift came from the north. In the southern hemisphere, however, it came from the opposite direction. In both cases its movement was from the pole toward the equator. The distance to which boulders have been carried is sometimes very great, even hundreds of miles. We know where they came from by their composition and other circumstances. Sometimes long trains of boulders are traced, as if an iceberg slowly sailed along, dropping continually as it went its rocky

freights. In such cases it is very easy to find the source of the boulders. Drift, in its passage, left in many places its marks, as scratches, and furrows, and smoothings. Sides of mountains exhibit often these marks, and so high up in this country that we know that there is only one mountain in New England, Mount Washington, that lifted its head above this drift action. I will not dwell longer on this interesting subject here, as it will be brought before you more particularly in another part of this book.

233. **Subsidences and Elevations of the Earth's Crust.**—You will find, as I proceed to notice the different ages of the earth's formation, that different portions of its crust were alternately submerged under the waters and elevated above them. This was not done by any sudden convulsive movements, but slowly, perhaps as slowly as the change of level now occurring in Sweden and some other countries, as noticed in § 206. These alternate subsidences and elevations, which were generally several, sometimes many in number, were necessary, as you will see, for the formation and arrangement of the various strata that make up the earth's crust. Sometimes subsidences have been prolonged through even many ages. Lyell speaks of one over a large area in England and Wales which continued so long that strata over six miles (32,000 feet) in thickness were deposited during the time. In all this period, covering not an age merely, but successive ages, there was an ocean's bed in that region, and this bed sank gradually and quietly as the deposits of solid matter were made upon it from the water lying above it. In no one of the ages of the earth's formation was there so remarkable an alternation of subsidences and elevations as in that one in which the coal was made and laid down. Each bed of coal, as you will learn more fully in another chapter, was made from vegetation grown upon the surface, and this being submerged by a subsidence, rock was formed over it.

Then, to make another bed of coal, there was an elevation for another growth, which was in its turn submerged, to be covered by a rocky deposit. Where there are many beds of coal there must have been many of these alternate movements.

234. **Pebbles, Sand, and Earth.**—In several different connections I have remarked upon the ways in which the rocks are worked up into pebbles, sand, and earth. You have seen that, while other agencies are at work, water, either directly or indirectly, is the grand agent—directly by its own erosive power, and indirectly by grinding fragments of rock, from the large to the minute, against each other. By these means chiefly has the soil all been prepared; and it is nothing but comminuted rock, with soluble matter from some of the rocks dissolved in the water diffused through it. It is true that when seeds are put into the soil thus prepared additions are made to it from the decay of vegetable matter, but this is a mere return of such matter to the mineral condition. All was originally mineral—earth, water, and air; and in plants and animals we have life acting only upon mineral matter, giving it, for the time being, new properties. It is a very important item in the construction of the earth for man that the soil should be thus prepared, by what may, in one sense, be considered a destructive process, but in another the putting the material of which the rocks are composed into a special form for a special purpose. If this were not done, vegetation and animal life could have flourished, at best, to but a scanty extent.

235. **Earth-worms and Ants.**—These animals, apparently so insignificant, are real geological workers, accomplishing in the aggregate great results in preparing the soil to produce food for man and beast. The earth-worm burrows in the earth, loosening it, and leaving his casts here and there. He is not confined to loose soils, but is ever encroaching upon hard spots, especially if

they be wet, as any gardener may see occasionally in the trodden walks. He does more than loosen the soil. He brings it up in a comminuted state, as the casts will show, leaving his burrows below to cave in by the weight of the earth above them. Some observations have been made which show that these operations are of great extent and importance. Mr. C. Darwin, who made many such observations, says that "although the notion may appear at first startling, it will be difficult to deny the probability that every particle of earth forming the bed from which the turf in old pasture-land springs has passed through the intestines of worms." If you look among the spires of grass you will find the casts of these worms scattered about, for they are always at work swallowing earth and disgorging it, either upon the surface or into their burrows. The tendency of water is to wash from the surface the finer portions of the soil, carrying it away or down into the ground, thus leaving the coarser parts on the surface; but the earth-worms can remedy this difficulty by bringing up the comminuted matter. Furnishing bait, then, for fishing is but a small part of the earth-worm's vocation.

The ants are also at work somewhat in the same way, choosing drier spots than the earth-worms do. You see them in multitudes on the dry garden walk, where they make their galleries underneath, from which they bring the materials for their little piles on the surface. They, by loosening the soil, help to produce the vegetation which you are continually at work to destroy; but, though operating against you in the walk, they are every where else doing a good work for you in your garden. They are especially useful in tropical climates, where they rapidly take to pieces, as we may say, the accumulations of dead vegetable substance, and mingle it with the soil. Dr. Livingstone, the great traveler, says of the labors of the ants in the forests of Africa, "These insects are the chief agents employed in forming a fertile soil.

But for their labors, the tropical forests, bad as they are now with fallen trees, would be a thousand times worse."*

236. **Corals.**—Coral animals, which are of the class called Polypes, are the most important organic or living agencies that have contributed to the construction of the earth. These animals are very small, most of them exceedingly so, being less than the size of a pin's head. They live in companies together, sometimes each a separate animal, and sometimes united together by a fleshy mass, making what is called a polypidom, or household of polypes. When separate, each sits like a cap on the summit of a column of carbonate of lime, and there takes in its food as it can catch it from the passing water. To this mineral column he is continually adding, and the process is a singular one. The animal is ever dying below and growing above; and as the column may be considered his skeleton, he may be said to be continually leaving dead skeleton below him as he grows. The process is essentially the same, though modified, in the case of the polypidom. In this work each animal does but little, but in the aggregate vast results are accomplished. In the ages that are past these little animals have made immense additions to the limestone in the earth's crust, and they are continuing their work now. The rocks which result from their work are not really built up by these animals. They furnish the material, which is broken up and changed into limestone, though there are imbedded in many of these rocks the coral forms. And the material is not furnished by these animals alone, for shell-fish and other animals abound wherever there are corals, and their remains add to the mate-

* I have noticed the earth-worms and ants in this chapter, rather than in the chapter on present changes, partly because they have undoubtedly been great geological workers in past ages, and partly because it is appropriate to consider their work in connection with the general view of the preparation of the soil for the use of man.

G 2

rial of the rocks. Especially is this the case with the shells.

237. **Coral Reefs.**—Along many shores there are ridges or reefs made by coral animals. The mode of their construction I will describe. These animals do not begin their work on the edge of the land, for they like clear water, which can not be had close to the shore. But there is a limit away from the land beyond which they will not work, and this limit depends on the depth of water. They can not live beyond a depth of about 100 feet, and most often they choose a depth of 20 or 30 feet to begin their operations. Where the water deepens rapidly from the shore, they are not as far away from it as where it deepens gradually. Coral animals are confined to certain climates. Almost all the shelving shores in tropical seas are lined by the reefs which they make. There is a remarkable exception in the case of the western coast of South America. Here there are no corals, while they abound on the eastern coast in the same latitudes. The reason is that the western coast has sweeping along it a cold current from the antarctic regions. Many of the coral reefs are not yet raised up to the surface, and these hidden reefs are very dangerous to navigation. Some reefs are of very great extent. Along the coast of New Holland there is one over 1000 miles in length, and for 350 miles in one part of it there is no passage through it. Reefs are spoken of as being of two kinds, fringing and barrier; the former being near the shore, and the latter at some distance from it, with a deep channel between. The fringing reefs are very apt, after a time, to become a part of the main land. The reason is that there is a continual washing of material into the space between the land and the reef, both from the land and the sea, chiefly the former. Indeed, Florida was all made in this way, reefs having continually formed age after age, and then joined to the main land. Many islands have been formed by coral animals. A large part

of the islands of the Polynesian Archipelago, as well as many of those of the Indian Ocean, came from this source.

238. **Atolls.**—This name is given to certain coral islands having a peculiar arrangement, one of which, Whitsunday Isle, is represented in Fig. 80. There is, as you

Fig. 80.

see, a strip of land inclosing a lake called a lagoon. In some atolls this strip of land is unbroken, while others have one or more openings, so that the lagoon forms a harbor. These atolls are of various shape, sometimes almost circular, sometimes long and narrow, having various bends and indentations in the margin of land. Their size varies from half a mile up to sixty miles across. The manner in which they are constructed can be explained by Figs. 81 and 82. We will suppose an

Fig. 81.

island around which the coral animals build reefs; $a\ a$, Fig. 81, being the reefs, and $b\ b$ the water between them and the island. But in the figure a mere point in the middle of the island is seen standing out of the water, although the bottom of the reefs is far below the lowest depth which coral animals reach. What is the explanation? It is supposed to be this. When these animals just began to make the reefs, the island stood far up out of the water; but as fast as the reefs were built up the island subsided, keeping the surface of the reefs all the time of the subsidence under the water. The rate of subsidence must have been very slow to correspond with the slow growth upward of the corals, which has been calculated to be only the one eighth of an inch in a year. In a completed atoll, the island having wholly disappeared, we have a state of things indicated in Fig. 82, $a\ a$ representing a section of the margin of land, and $b\ b$

Fig. 82.

the lagoon. An atoll, then, may be looked upon as "the tomb and monument of an island altogether buried beneath the waves."

There are groups of these singular islands, in some cases extending over large spaces. In the Pacific Ocean there is a band of such groups 4500 miles long, and varying from 200 to 600 miles in breadth.

Many atolls rise up from a depth of 2000 feet, and therefore are mountains of coral rock standing in the water over a third of a mile in height. At the rate of an eighth of an inch a year, it took the coral animals 192,000 years to build such mountains.

239. **Calcareous Shells.**—I have said that in the formation of the coral rocks there were contributions of shells. In fact, they every where contribute material for the limestones; but in this case, as well as in others, the great work of furnishing material is done by very small, even minute animals. The shells of such animals have formed extensive strata in various parts of the earth's crust. In Fig. 83 you see represented the shells of some of the foraminifera, a class of these minute animals, so named because their chambers communicate by numerous foramens or pores. The microscopic animals

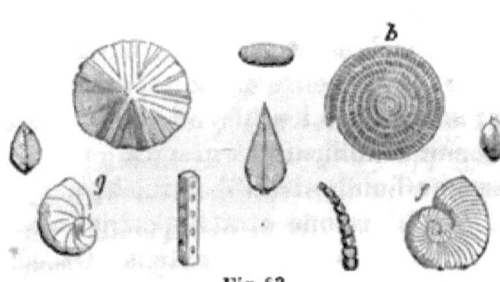

Fig. 83.

that inhabit these many-chambered shells are composed of a gelatinous fleshy substance, and have minute prolongations which they can throw out and retract, and which they use for swimming, crawling, and gathering their food. There is great variety in the shells of the different species, and the regularity, beauty, and delicacy of structure of these shells are very wonderful. In view of the vast amount of rocky material which such minute animals have added in past ages, and are now adding, to the earth's crust, it may well be said that the additions made by the remains of the larger animals, such as elephants, lions, crocodiles, whales, etc., are utterly insignificant in the comparison. There is one division of the foraminifera that are not microscopic, viz., the nummulites, one of which you see at *b*. They are of various sizes, from very minute up to the size of an inch and a half in diameter. The name comes from the Latin word *nummus*, money, because the shell resembles a coin in shape. Sometimes limestone is composed entirely, or nearly so, of nummulites, and then is called nummulitic limestone. The Sphinx and the Pyramids are made of this rock. It constitutes the principal part of several mountain ranges in the south of Europe, as the Alps and Pyrenees.

In soundings in different parts of the Atlantic Ocean, between Ireland and Newfoundland, as far south as the Azores, there has been brought up a soft, sticky mud, which has been called *ooze*. This, on being dried and examined with the microscope, was found to consist of the minute shells of foraminifera. As these shells are composed of carbonate of lime, when an acid, as the sulphuric, is added to ooze, brisk effervescence occurs, because the acid, taking the lime, sets the carbonic acid gas free. Here there is an immense chalk deposit going on over a large area; and in such lengths of time as were occupied by deposits and solidifications in ages gone by, this deposit may become hundreds of feet thick, and be solidified into rock. In limestone strata quarried near Paris the rock is composed to a large extent of shells no larger than millet-seeds. Ehrenberg has discovered in that form of limestone which we call chalk, animals much smaller than this—so small that, where they nearly constitute the whole mass, as they do in the chalk of southern Europe, there are over a million of them in every cubic inch. Soldani collected from less than an ounce and a half of rock from the hills of Casciana, in Tuscany, 10,454 shells of foraminifera of various species. Some of the species are so minute that it would require over a thousand of the shells to weigh a grain.

240. **Silicious Shells.**—You have already learned in § 116 that silica, or flint, is dissolved by certain means in water, and then is gathered up by plants and animals, becoming again solid in soluble silica. There are very extensively scattered in the earth innumerable minute animals and vegetables which thus gather this material from the water in which they live, and then, as they die, their silicious shields or shells are deposited, and much of the deposit becomes in time solid rock. These shields present great variety of figure, and are very beautiful as seen under the microscope. In Fig. 84 you see represented three of the forms of shells of this kind, of which a stratum of white clay about Richmond, Va., is chiefly composed. The stratum is from twelve to thirty feet thick, and is of great extent. The tripoli, or rot-

Gaillonella sulcata.

Actinocyclus.
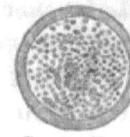
Coscinodiscus patina.
Fig. 84.

ten-stone of Germany, forming beds sometimes fourteen feet thick, is made up mostly of silicious shells so minute that a cubic inch contains forty-one thousand million (41,000,000,000), the weight of a single shell being calculated to weigh the $\frac{1}{180}$ millionth part of a grain. Silicious deposits from vegetables and animals are often mingled with calcareous deposits, and hence came for the most part the flint-stones, which, as you will see hereafter, are so often found in chalk. One tenth of the oaze (§ 239) is silicious. So much agency are the minute silicious animalcules and plants known to have had in furnishing material for silicious rocks, that some have supposed that most rocks of this kind have received their material from this source.

I remark in passing that, although the diatoms of which some forms are given in Fig. 84 have heretofore always been considered as animalculæ, Professor Dana states that they are now regarded as vegetable.

241. **Seas, Lakes, and Rivers.**—I have thus shown you what means have been employed in the building up of that part of the earth which is included under the general term land. The great cavities of the sea, "the store-houses" in which God "layeth up the depth," we know less about, and our researches are confined mostly to the comparatively shallow portions of the ocean which skirt the land. The manner in which the rivers that run into the sea are made I need not stop to point out to you. The streamlets made by every shower illustrate that subject, so that, with what has already been said in regard to rivers, your own observation can give you a proper insight into it. The same can be said, for the most part, of the lakes.

242. **Plan in the Construction of the Earth.**—Enough has been developed in what I have said of the construction of the earth to show you that, although there have been commotion and much apparent confusion, there has been a well-ordered plan throughout all the changes,

aiming at the eventual fitness of the earth for man as his habitation. This will be much more obvious to you as we proceed to notice the different ages of the earth's preparation, and show how the continents have grown to be what they are by successive additions. In the case of each continent, **the form of the land first lifted above the surface of the waters had a manifest reference to that form which was to be given to it when it was to be completed and made fit for the use of man.** There was in the beginning a germ, and the gradual unfolding of it has been beautifully traced out by the labors of the geologist.

## CHAPTER XII.

### RECORD OF LIFE IN THE ROCKS.

243. **Life in the Different Ages of the Earth.**—There was a period, and that a very long one, as you will see in the next chapter, in which there was no life upon the earth. But after that period was passed, life, vegetable and animal, was introduced, and became, as you have already seen, a great agency in the construction of the earth's crust. What the forms of life were in the succession of ages previous to the age of man have been made out by the geologist, by observing the remains of vegetables and animals found in the strata of the rocks. The life of the world, then, may be said to have written its own history on tables of stone, and we are able to read on these tables the various changes which the forms and modes of life have taken on during the ages of the past. The lengths of these successive ages we can not make out with any accuracy; we can not measure time in the far past of our earth by years and centuries as we can in the present age; but we can, by the life-record, learn the order of succession, which has marked both the construction of different parts of the earth and the furnish-

ing of it with the forms of life. In the successive ages there have been variations in the vegetables and animals, especially the latter, which designate these ages, as we learn by reading the record of the rocks. The farther we go back in this record the less is the resemblance of the forms of life in the past to those of the present, and the gradual increase of this resemblance, as we come down from the first dawning of life on the earth to the present age, is very obvious. This great fact will be developed to you in various ways as we proceed.

244. **Nature of the Evidence.**—You have already had some glimpses of the record, and see what is the nature of the evidence that it affords. You have seen that the strata were formed one upon another in a certain order, and that, as the material of which they are composed was deposited, various remains of vegetables and animals became mingled with it, and in the solidification made a part of the rock. You see, then, how, in the midst of the various disturbances of the strata, the geologist can determine the relative ages of rocks by examining the remains of organic substances contained in them. The strata may be vertical, or may even be so far bent over that the older rock may lie upon the more recent, but the life-record reveals the truth; and, farther than this, the record is essentially the same in different regions of the earth, so that the conclusions of geologists in one region may be applied in another. For example, coal is found in different countries in connection with strata that contain certain organic remains; for, in the construction of the earth, there were certain periods for the formation of coal, and one especial period for this purpose. In whatever country, then, such strata are found, it may be rationally expected that coal may be found in connection with them. Other examples might be given of a similar character, but this will suffice.

245. **Fossils.**—Any remains of any kind, of vegetables or animals, found in rocks or in loose earth, are called

fossils. The term comes from the Latin word which means to dig. The remains are generally more or less defective, the softer parts being removed, and only portions of those which are hard being preserved, although in some cases there has been entire preservation, as in the case of ancient mammoths found imbedded in frozen earth in Siberia. Sometimes a fossil is a mere trace, as of a leaf, or simply a track of an animal. Sometimes in the fossil there is not a particle of the original organized substance, as when perfect petrifaction takes place (§ 118). The mineral substance in these petrified fossils differs in different cases. The most common substances are silica, carbonate of lime, and clay. Analogous to petrifaction is the substitution of some mineral substance for the whole of an organic body, as a shell, the new substance making merely a cast of it. This differs from a petrifaction in that there is no trace of the texture of the shell in the cast. Coal, strictly speaking, is a fossil, for it is the remains of the vegetable substances from which it came, their texture being retained in it more or less, as shown in § 41.

There is much difference between different organized substances in regard to their preservation as fossils. Bones are very durable, but the bones of birds are seldom found, because they were borne up by their feathers as they decayed, and the bones, being hollow, were easily broken. Their tracks, however, have been extensively observed in the layers of rock, as you will see farther on, and the investigation of these, especially in this country, by Hitchcock and others, has furnished one of the most interesting chapters in geology. Insects, too, are seldom found, because they are so light, and decay so rapidly from the presence of air in the air tubes that pervade their bodies. Such hard substances as shells are more largely preserved than any other substances.

246. **Abundance of Fossils.**—Although many organized substances, from the softness of their structure,

could not be preserved at all, and the hard structures were liable to be broken up or even destroyed, yet the rocks abound in fossils. It is stated by Agassiz and Gould that in most formations the number of species of animals and plants found in any locality is not below that of the species now living, in an area of equal extent and of a similar character. For example, a coarse limestone in the neighborhood of Paris contains not less than 1200 species of shells, more than twice the living species now found in the Mediterranean. So, too, in a certain kind of limestone in New York, called the Trenton limestone, there have been found 170 species of shells, nearly as many as are now found on the coast of Massachusetts. The same is true of the individual plants and animals as of their species. Extensive strata, as you have already seen, are formed of the remains of animals, as corals and shells, and the immense stores of coal are made up of the remains of plants. Then there are the immense quantities of silicious fossils, for so they may be called, furnished chiefly by those minute plants called diatoms, noticed in § 240. Of these the microscope shows us that there is a great variety of species.

247. **Mode of Investigating Fossils.**—It is on the principles of the science of comparative anatomy that the fossils are investigated. This science may be said almost to have been founded by Cuvier, who acquired a marvelous skill in prosecuting its researches. From a few bones, sometimes even one small bone, of an unknown animal, he could make out the construction of the whole frame, and the character and habits of the animal. The principles upon which this is done are very simple. As in a machine each part has a relation to every other part, so it is in the machinery of an animal; the relations in this case, however, being more perfect, because the builder of the machine has perfect wisdom. For example, teeth of a certain kind not only require a certain kind of jaw, but a certain kind of feet also. An

animal having the sharp, tearing teeth of a beast of prey would starve if he had such feet as grass-eating animals have. These principles almost every one applies, to a certain extent, with great ease. The sight of a tooth, or claw, or beak will suggest at once to our minds the habits and the shape of the animal to which it belongs. And by careful observation one can acquire great skill in applying these principles to the minute parts of animal frame-work, so that a familiar acquaintance can be formed with the relations of each individual bone. The same thing is true, to a certain extent, of vegetables. Now as in fossils there are often only parts of a vegetable or animal, this skill of which I have spoken is brought into requisition, and by it the living beings of the past are marshaled into classes, genera, and species, just as the beings of the present are. Palæontology is the name which is given to this science of the fossils. It is a word derived from three Greek words—*palaios*, ancient; *ontos*, the genitive of the word for being; and *logos*, discourse. It is really the application of two sciences, botany and zoology, to the remains of life left by the past. Some, however, use the term in a more restricted sense, considering it as meaning the science of fossil *animals*, while the term Fossil Botany is applied to the science of fossil plants.

248. **Living Structures of Former Ages like those of the Present.**—Vegetables and animals were constructed on the same fundamental plans in the far past that they are now. The same general divisions existed. And the differences result from variations which were made to meet different circumstances, and not from any essential alterations in general plans. This being the case, it is appropriate, before we go farther, to look at the outlines of the Creator's plans in the construction of the forms of vegetable and animal life. A knowledge of them will aid you materially in following out the developments which I shall make to you in the succeeding chapters, in

249. **Amphigens.**—The vegetable world presents to us five groups, marked by different modes of growth. First we have the amphigens, so named because they increase by growth on all sides, the name being derived from two Greek words, *amphi*, around, and *gennao*, to produce. This mode of growth can be seen in the lichens that spread out upon the surfaces of rocks, and trunks of trees, and old fences. This division includes, besides these lichens, the scum-like growths that you see on stagnant water (Confervæ), the mushroom family (Fungi), and the sea-weeds (Algæ) that lie upon the shelves and ledges of the shallow places in the sea. These are all represented in Fig. 85. They are the lowest kinds of

Fig. 85.

vegetation, and are found where no other plants can grow. They may be considered as pioneers of the higher orders of plants, helping to weather the rock, and to gather upon it the material for the growth, in time, of other plants, and enriching also barren soils by their decay, so as to prepare them for higher forms of vegetation. These plants are wholly cellular in their composition, while all other plants, as grasses, trees, and shrubs, have tubular vessels for the circulation of the sap. The latter are called, therefore, *vascular* plants, and the former *cellular*.

250. **Acrogens.**—This division includes the mosses, the ferns, the horsetails, and the ground-pine family. These plants are pictured in Fig. 86, the tall tree-fern that grows in tropical climates being represented as well as the common fern, or brake, that is so familiar to us in temperate climates. The plants of this class delight in swamps and shady places. Their remains are found in the peat of the present day, and in the coal deposits of past ages. Their title comes from the fact that they increase from the top alone, it being derived from two Greek works, *akros*, summit, and *gennao*. These, like the amphigens, are of service in preparing soil for the nourishment of higher orders of plants. They do a great work in this respect, because they have a rapid growth, and from year to year, by their decay, add to the solid, nutritious material in the swamps and damp places where they so luxuriantly flourish. Neither the amphigens nor the acrogens afford support to animal life to any extent.

251. **Gymnogens.**—In this group, Fig. 87 (p. 168), we have the cycads, or pineapple family, and the first pines, or cone-bearing trees (coniferæ). They have this name (*gymnos*, naked, and *gennao*) because their seeds are naked instead of being inclosed in cases. They appear in various kinds of soil, the dry as well as the damp. From them, like the acrogens, peat and coal have been largely formed. Our coal, both bituminous and anthracite, was laid down in beds from the growth of both acrogens and gymnogens far back in the past, as you will see in another chapter. Their stiff, juiceless leaves, scaly seeds, and hard berries afford but little nutriment

Fig. 86.

for animals, and therefore the record of the rocks shows but scanty remains of any of the higher orders of animals amid those of these plants. The inference is a clear one that when, in the ages past, these vegetables were

Fig. 87.

made to flourish for special purposes, but few of the animals of higher orders were brought upon the scene, because there was little food which such animals could appropriate to themselves. The principle of adaptation was applied to the mutual relations of the vegetable and animal worlds in those ages as thoroughly as now by an all-wise and omnipotent Creator.

252. **Endogens.**—In this division of plants we have the grasses, rushes, lilies, canes, and palms, as exhibited in Fig. 88. Here we have a decided advance on the pre-

Fig. 88.

viously mentioned groups in the objects which the Creator had in view in introducing them. While the plants of this class, like those of the former classes, add by their decay to the nutritious material of the soil, their chief design is to afford nutrition directly to the animal kingdom. What was in the other classes only an incidental object, is in this the primary one, while the primary of those classes is here the incidental. The chief object of

H

the amphigens, gymnogens, and acrogens was to build up, as we may say, the soil of the earth, and of a part of them to furnish the coal which has been stored up for the use of man. They are *geological* in their agency—that is, earth-making. But the chief object of the endogens has been to furnish food to animals—that is, animal-making. They are *zoological* in their agency. The name endogens comes from *endon*, within, and *gennao*, and expresses the fact that the growth is by addition in all the parts of the plant equally. In the trunk or stem, for example, there is in every addition a formation of new fibres intermingled with those already present, and in old trunks or stems the hardest part of the wood is toward the surface, and the softest toward the centre. Endogenous plants have no pith and no distinct bark.

253. **Exogens.**—This name comes from *ex*, out, and *gennao*. It indicates an entirely different mode of growth from that of the endogens. There is a pith and distinct bark, and the addition is in rings outside of those which are already formed, the hardest part being toward the centre, and the softest toward the surface. The difference in the modes of growth in the endogens and exogens is exhibited by sections of stems or trunks in Figs. 89 and 90. In the endogenous stem, Fig. 89, you see the holes occasioned by the section of the circulating vessels, and between these is the mass of woody or cellular tissue which makes up the substance of the stem, and forms, indeed, the walls of the vessels themselves. But in the exogenous stem you see the pith, the rings of wood, and the bark. The lines which radiate like the spokes of a wheel from the pith to the bark are called

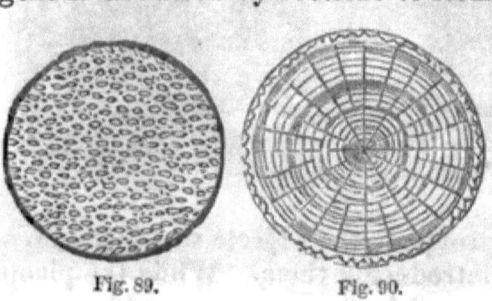

Fig. 89.   Fig. 90.

*medullary rays*, and indicate plates of cellular or woody substance. These make what is called the silver grain of the wood. This division of plants includes the herbs, shrubs, and timber trees, as they are seen grouped in Fig. 91. In this and Fig. 88 you see the differences be-

Fig. 91.

tween the two classes of plants exhibited in the trunks of two trees represented as cut square off. The same can be said of the exogens as was said of the endogens in relation to their *zoological* character. The remains of few of either class are found among the fossils of the far past, but the record of life in the rocks shows that they were more and more introduced the nearer we come down to the age of man in tracing the successive ages.

254. **Investigation of Vegetable Fossils.**—We can not study fossil botany in the same manner that we do the botany of the present. In the latter case we have all the parts of plants in full, but in the former merely remains, and sometimes only scanty ones. In many cases, therefore, in fossil botany we can only approximate to a true classification. With the scanty evidence often presented, it is not surprising that some mistakes should have been made. One of these I will mention. For a long time certain trunks of trees found in the coal measures were put in a different family from certain roots found also in these measures; but at length it was discovered that the roots and the trunks belonged to the same tree, though in the strata they were found separated from each other. Of the many points of distinction in regard to vegetable fossils I will notice but one, the venation of leaves, or the mode of distribution of their veins. The veins in the leaves of endogenous plants run parallel with each other, as seen in Fig. 92, while in the exogenous the veins interlace, as seen in Fig. 93, which represents the leaf of the apple.

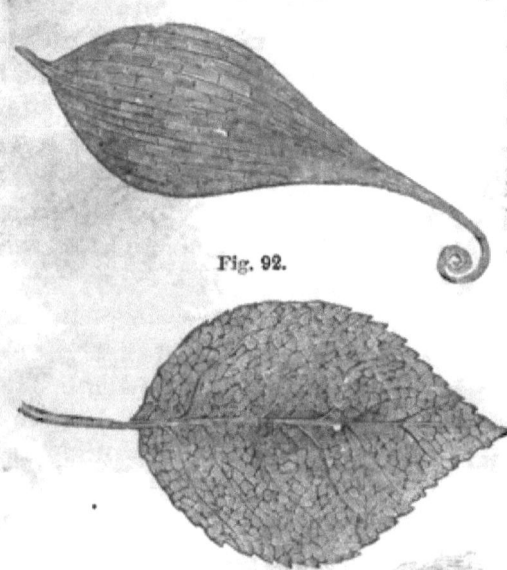

Fig. 92.

Fig. 93.

**255. Animal Fossils.**—The study of fossil zoology is more satisfactory than that of fossil botany, for three reasons. First, the distinctions between different kinds of animals are more readily made out from the partial remains which the rocks present to us. The reason is that the differences between animals are of a more decided character, and especially those which mark the grand divisions into classes, as you will soon see. Secondly, most animal remains are in a better state of preservation than vegetable remains. Shells, corals, scales, teeth, and bones are very durable. Thirdly, more attention has been paid to fossil zoology than to fossil botany by geological observers.

**256. Protozoans.**—This term, derived from two Greek words, *protos*, first, and *zoon*, animal, is applied to a class of organisms which seem to stand on doubtful ground, being half vegetable and half animal in their characteristics. They are the sponges, the foraminifera, and the infusoria. They have, for the most part, an indistinct organization, so that they can not be classified with any of the great divisions of the animal kingdom. Professor Dana therefore speaks of them as systemless; that is, as belonging to no system. And yet it would seem as if some of them have a sufficiently definite form to be classified, as operculina, *f*, polymorphina, *g*, and nummulites, *b*, in Fig. 83, which are certainly very much like ordinary shell-fish, though their resemblance in certain respects to polypes may render it improper to locate them with shell-fish in a definite classification. The sponges have been thought by some to be on doubtful ground, but it is now considered settled that they are animal. The flinty diatoms, on the other hand, as stated in § 240, are now to be placed among vegetable organisms, and are not, therefore, any longer to be called animalculæ.

I pass now to the consideration of those animals that can be definitely classed in systems.

**257. Radiates.**—This class includes the coral animals, hydras, sea-anemones, sea-nettles, echini, or sea-urchins, and star-fishes. The name comes from the ray-like arrangement which is so manifest in many of these animals, and exists really in all. It is from this radiate form chiefly that many of them were formerly supposed to be plants, and even now they are often called plant-animals. Examples of them are seen in Fig. 94. Here

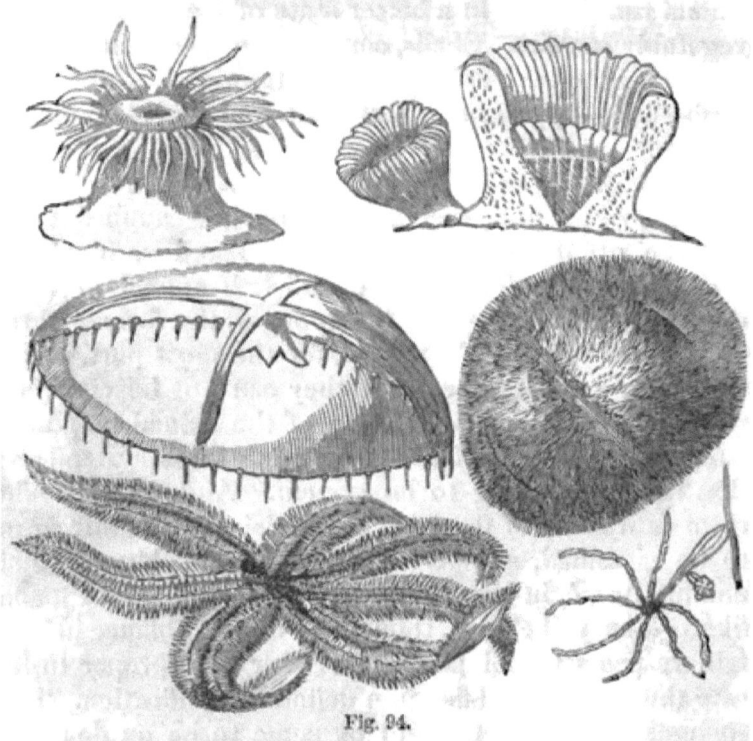

Fig. 94.

are represented the star-fish, the little hydra, a medusa, a sea-urchin, and sea-anemones. One of the anemones is divided in half, that you may see the interior. They are all inhabitants of the ocean, and many of them, as they are seen sometimes clustered together over a considerable space, present, with their beautiful and varied colors, the appearance of a garden under water. They are pulpy in their consistence, many of them, however, having a hard frame-work, as the coral animals and the starfishes. They live to a great extent on the microscopic organisms which abound in the sea; and, though they are more *zoological* than the protozoans, they are largely *geological*, especially the coral animals, which have contributed, and still contribute, as you have seen (§ 236), such quantities of limestone to the earth-making processes.

258. **Mollusca.**—This class, Fig. 95, contains animals

Fig. 95.

of a much more complicated and varied character than the preceding. They are not confined to the ocean, but many live in lakes, rivers, and swamps, and some even on the dry land. The substance of these animals is soft, as seen in the common oyster, and hence the name, which comes from the Latin *mollis*, soft. But most of them have hard shells, and so are called, in common language, shell-fish; though some, as the slugs, are wholly uncovered. This class includes oysters, mussels, clams, snails, slugs, nautili, cuttle-fishes, etc. The shells of some of these animals have contributed largely to the formation of the limestone of the earth, but they are more zoological in their relations than the previous class. So widely distributed have shell-fish always been, and so durable are their shells, that their fossils are of great use to the geologist in determining the relative ages of the rocks.

259. **Articulata.**—This sub-kingdom includes insects, worms, the spider and scorpion tribe, and the crab tribe. These varieties are represented in Fig. 96. You have here animals of extremely various character in many respects. Nothing, for example, could be more unlike than a butterfly and a lobster. Yet all these varieties agree in one thing—in having a covering which answers to them as a skeleton, in giving firmness to the body, and furnishing points of attachment to their muscles. This is very obvious in such animals as crabs and lobsters, but it is equally true of insects and worms. This skeleton coat of mail, as it may be called, has commonly a very manifest ring-like arrangement; and, as the rings are jointed with each other, the name articulata, coming from the Latin *articulus*, joint, is given to this division of the animal kingdom. The articulata are mostly zoological in their relations, and are only slightly geological. They live in every element—in the air, the earth, and the water. Some are animal-eaters, some vegetable-eaters, and some both, and they, in their turn, are the food of other animals of various kinds.

260. **Vertebrata.**—The animals of this sub-kingdom have an internal skeleton with a back-bone, so called in

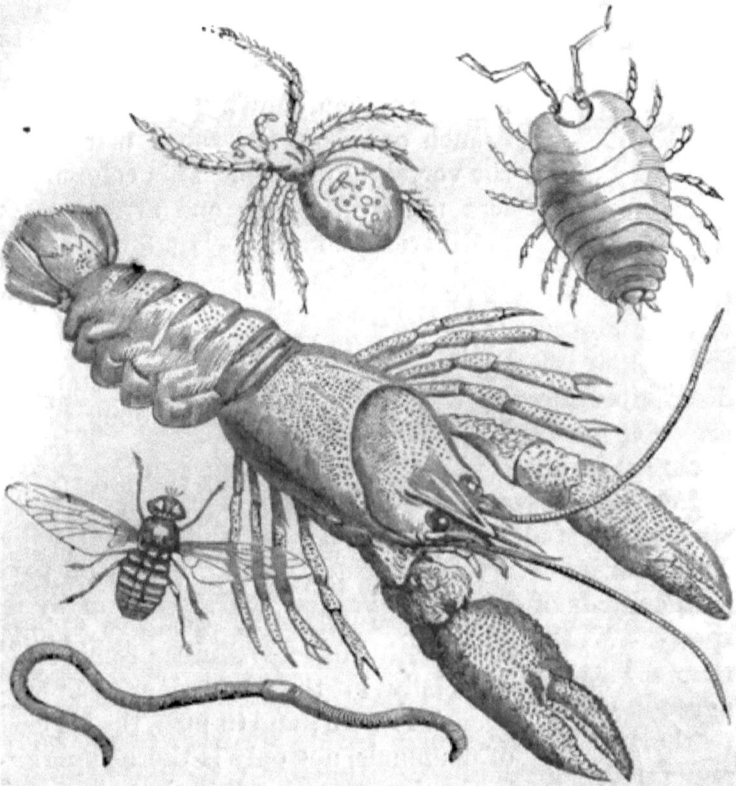

Fig. 96.

common language, made up of vertebræ.* Through this chain of bones runs a nervous extension of the brain called the spinal marrow, from which nerves branch out to different parts of the body. In Fig. 97 you see represented a single vertebra from the spinal column or back-bone of man, *a* being its front part, and *b* its pointed rear part, which you feel on tracing the line of the

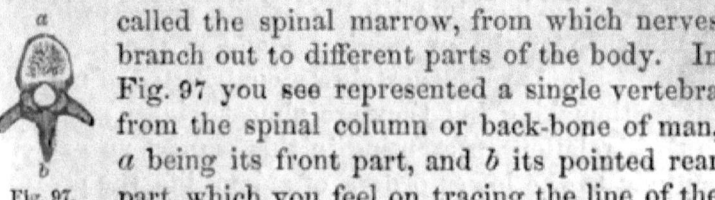
Fig. 97.

\* Sometimes animals are spoken of as being in two great classes, the Vertebrates and the Invertebrates, those which have and those which have not vertebræ, the prefix *in* being a particle of negation. The Invertebrates include the three great classes already spoken of— Radiates, Mollusks, and Articulates.

Fig. 98.

back-bone. You see the opening which is filled with the spinal marrow. In Fig. 98 you see the brain and spinal marrow of man, showing how the nerves branch out from the spinal marrow in the vertebral column. This column differs much in length and arrangement in different animals, being, for instance, vastly longer in the serpent than in man, and having a number of vertebræ in proportion to its length. The vertebrata have four grand divisions — mammals, or those that suckle their young, birds, reptiles, and fishes. These are represented in Fig. 99. For the most part, it is in this sub-kingdom that we have the greatest complication of structure, and with it the highest manifestations of life. Intellectual qualities appear here with their greatest prominence, especially in those portions that approach to man; and in man, the highest of mammals, not only is there superiority in degree of intellect, but there are superadded powers, making his intellect different in kind as well as in degree, thus linking him with the Infinite, and showing the significance of the expression, made in the image of God. Little have the vertebrata contributed of material for earth-making, but they are almost wholly zoological in their relations, presenting in this respect an extreme contrast to the lower orders of animal life, the protozoans and the radiates. The relations of the vertebrates and the higher orders of the vegetable world, the endogens and the exogens, are very obvious, and accordingly we find in the life-record of the rocks the evidence of their introduction together upon the world's arena, the

Fig. 99.

full introduction of both being reserved for the age which ushered in the advent of man.

**261. Divisions of the Earth's History.**—As the record of the rocks is a life-record, it seems eminently proper to base the division into ages upon the changes which we find in the forms of life from age to age. There are other modes of division; but, as the developments of the life-record have come out, the tendency has been more and more to a division on this basis. This division has been made differently by different authors. The one which I shall follow is that which is brought out in that grand American work, Dana's Geology. He divides the geological history of the world into seven periods: the Azoic age, in which there was no life (*a* privative, and *zoe*, life); the age of Mollusks; the age of Fishes; the age of Coal-plants, or the Carboniferous age; the age of Reptiles; the age of Mammals; and the age of Man.

The last age is still in progress; the others are in the past. How long they were we know not; but that they were each much longer than the age of man has thus far been we have the most conclusive evidence to show.

It is proper to notice here a division of time into five periods which is common in works on Geology, and in relation to which there are certain terms that are in constant use by all geologists. 1. *Azoic* time. This has been already explained. 2. *Palæozoic* time. This was a period in which the forms of life, as their remains show, were very ancient—that is, differing decidedly from the present. This period includes the three ages of mollusks, of fishes, and of coal-plants. Its name comes from *palaios*, old, and *zoe*. 3. *Mesozoic* time, the fossils of which differ less than those of the Palæozoic from the living forms of the present, the term being derived from *mesos*, middle, and *zoe*. The ages included in this period are the middle ages of geological history, and cover what is called the age of Reptiles, which is really, like all the other grand divisions of geological time, a succession of long ages. 4. *Cainozoic* (or, as it is sometimes called, Cenozoic) time, from *kainos*, recent, and *zoe*. This is the age of Mammals. 5. The *Present* age. The terms primary, secondary, and tertiary are used often as meaning the same, respectively, as Palæozoic, Mesozoic, and Cainozoic.

**262. Boundaries of the Ages.**—No age has had a sharply-defined limit in relation to its life-record, but each age has been to some extent mingled with other ages—in its rise with the age that preceded it, and in its decline with the age that followed it. Geological history, as Professor Dana remarks, is like human history in this respect, ages in both cases being marked by peculiarities which, indistinct at first and foreshadowed in a previous age, at length come out prominently to view. When they thus culminate it is easy to trace them back to their rise, and follow their growth, and then their decline, as they lose themselves in the rising peculiarities of another coming period or age. Thus that remarkable age, the coal-bearing or Carboniferous age, which was specially occupied in making and storing up coal in the strata of the earth's crust, was foreshadowed by the occurrence of plants similar to the coal-plants in the previous age, the age of Fishes. So also mammals appeared to some extent long before that age in which these animals were so numerous, in many cases so monstrous, as to make it proper to denominate it the age of Mammals. It may be remarked, in this connection, that it is not the idea to give the name of a class of animals to an age because they are more abundant then than they are in any other age. Thus the mollusks are really not as abundant in the age to which they give their name as they are afterward in the age of Reptiles, but in the age of Mollusks they are more abundant than any other class. At the same time, abundant as are the mollusks in the age of Reptiles, the latter then surpass them in abundance, and so give the name to the age. The articulates, which do not give a name to any age, have steadily increased from their small beginning far back at the conclusion of the Azoic age on to the age of the present.

## CHAPTER XIII.

### AZOIC AGE.

**263. Beginning of Solidification of the Earth.**—There was a time when the earth was a liquid mass. Of course, with such a degree of heat as sufficed to maintain it in this condition, there was no liquid water, but the heated ball was enveloped in steam. But after a time a portion of the outside of the mass became solid, forming a crust. Professor Hitchcock says that it is not unlikely that the time occupied in cooling the earth from its melted state, sufficiently to form a crust, was longer than all the time of the ages which were occupied in laying down the strata that contain fossils—that is, longer than all palæozoic, mesozoic, and cainozoic time. And if the earth, as is supposed by many geologists, was once so heated as to be in a vaporous state, the time of cooling must have been vastly longer than this. But, whether we reckon the azoic or lifeless age of the earth as beginning with the first solidification of the crust, or with the melted state, or extend it back to the supposed vaporous or *nebulous* state, so called, the time that elapsed before any life appeared on the earth was immense in length.

264. **Floor of the Earth's Crust.**—It was in the solidifications of the Azoic age that the floor, as we may term it, of the earth's crust was laid. This floor lies over the melted matter which is inclosed still within the crust, and which constitutes by far most of the bulk of the globe. It is not an even floor, although it was laid down in horizontal strata. It has been bent, and folded, and fractured, and lifted up, and broken through by melted masses forced upward from below. Although, then, its strata were laid before any of the fossiliferous strata, yet, from these bendings and upheavals, they appear in some parts of the earth upon the surface of its crust, the fossiliferous rocks skirting their prominences or partially overlying them.

265. **Changes in the Azoic Rocks.**—Vast changes occurred in the azoic rocks during both the Azoic age and the succeeding ages, for fire and water were at work upon them more or less during all that time. Just as soon as solidification took place on the surface of the great melted ball which once constituted our earth, a large part of the steam surrounding it was condensed into water, which of course fell in rain. At the same time, the forming crust, as is the case with all matter except water, in passing from the melted to the solid state,

contracted, and this contraction crumpled it into folds, regular and irregular, and thus made channels and cavities for the water to run and dash in. Water then began that great work of denudation, which, as you have seen in previous parts of this book, it has been carrying on ever since. From that denudation in this first age of the world was supplied material for the strata of the grand azoic floor, which covers up the fiery deeps within, and upon and against which the rocks of after ages were laid. So great have been the changes in the azoic rocks, that it is impossible to distinguish among them any that can with certainty be made out to be portions of the original crust—that is, portions of the first solidification.

266. **Some of the present Land formed in the Azoic Age.**—In some quarters of the earth the azoic rocks appear upon the surface of the earth's crust. These, as Professor Dana states, are "either, 1. Those which have always remained uncovered; 2. Those which have been covered by later strata, but from which these superimposed beds have been simply washed away without much disturbance; 3. Those once covered, like the last, but which, in the course of the upturnings of mountain-making, have been thrust upward among the displaced strata, and in this way have been brought out to light." The areas in which the rocks appeared above the surface of the water during the Azoic age were scanty compared with the whole area occupied by the land at the present time. These areas were the beginnings of the continents, and it was by successive additions to them during the succeeding ages that the continents grew to be what they now are.

267. **America and Europe in the Azoic Age.**—The azoic rocks that come to the surface in North America occupy, for the most part, a very long and comparatively narrow strip of land extending from Nova Scotia west to the base of the Rocky Mountains. This monstrous island

was not straight, but bent upward either way, forming an elbow, and shaped much like Fig. 100. If you com-

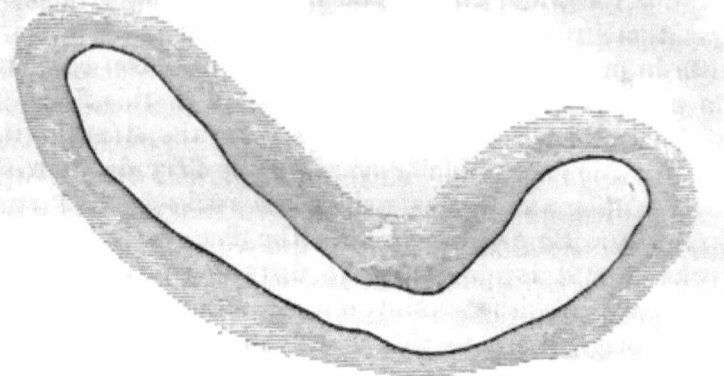

Fig. 100.

pare this with the shape of North America as a whole, you will see why this beginning of the continent had this shape, as it was completed afterward by additions. The elbow extends downward, as the lower part of the complete continent does, and the larger limb of the island corresponds with the more extended side of the continent. Some few other comparatively small azoic areas are found in North America, one of which is in Missouri, and includes the noted Iron Mountains. All else was sea in that quarter to the end of the Azoic age. Agassiz says of the long island which was the beginning of the North American continent, "We may still walk along its ridge, and know that we tread upon the ancient granite that first divided the waters into a northern and southern ocean; and, if our imaginations will carry us so far, we may look down toward its base, and fancy how the sea washed against this earliest shore of a lifeless world."

Europe was quite in contrast with America in this age. There was no one great island, but several of considerable size, and some small ones were scattered about in that part of the vast azoic ocean. There was no so obviously marked beginning of a continent as in the case of North

America. It is with great truth, then, that Agassiz says that, though America is called the New World, she is really the Old World, for she is really first-born among continents.

**268. Thickness of the Azoic Strata.**—The strata of the azoic rocks varies from twenty to thirty thousand feet in thickness. An immensely long period was required to deposit and solidify so much material, and comparative quiet must have reigned during their deposit. The upheavals occurred after this was done.

**269. Upheavals and Bendings.**—These are very extensive, and there is much regularity in them, for the strike is the same in some cases over great extents of territory. The regularity is most observable in the strata of the latter part of the age, for in the first part the heat was so great that the rocks then formed are much contorted.

**270. Rocks of the Azoic Age.**—The rocks of this age are granite, gneiss, schists, limestones, etc. They were mostly deposited in beds, and then were crystallized chiefly by the influence of the great heat which prevailed in that age. Some of the granite is metamorphic, but some of it was made originally as granite, having been forced upward from below, as was often done at different periods in the succeeding ages. One characteristic of the azoic rocks is the prevalence of iron. Some of the minerals which enter into the composition of some of the rocks contain iron, and there are sometimes found in the strata beds of iron ore of greater extent than in any other age.

**271. Heat and Light in the Azoic Age.**—The heat of the forming crust in the first part of this age must have been very great, and the upheavings and commotions must then have been tumultuous. Hugh Miller thus describes what may be imagined to be the state of things at that time. "Let us suppose that during the earlier part of this period of excessive heat the waters of the ocean had stood at the boiling point even at the surface, and much higher in the profounder depths; and farther, that the half-molten crust of the earth, stretched out over a molten abyss, was so thin that it could not support, save for a short time, after some convulsion, even a small island above the sea level. What, in such circumstances, would be the aspect of the scene, optically exhibited from some point in space elevated a few hundred yards over the sea? It would be simply a blank, in which the intensest glow of fire would fail to be seen at a few yards' distance. An inconsiderable escape of steam from the safety-valve of a railway engine forms so thick a screen that, as it lingers for a moment, in the passing, opposite the carriage windows, the passengers fail to discern through it the landscape beyond. A continuous stratum of steam, then, that attained to the height of even our present atmosphere, would wrap up the earth in a darkness gross and palpable as that of Egypt of old—a darkness through which even a single ray of light would fail to penetrate. And beneath this thick canopy the unseen deep would literally "boil as a pot" wildly tempested below; while from time to time, more deeply seated, would upheave suddenly to the surface vast tracts of semi-molten rock, soon again to disappear, and from which waves of bulk enormous would roll outward, to meet in wild conflict with the giant waves of other convulsions, or to return to hiss and sputter against the intensely-heated and fast-foundering mass, whose violent upheaval had first elevated and sent them abroad." At length, however, the earth's forming crust would cool down, so that the steam atmosphere would become less thick, and after a time the rays of the sun would struggle through, forming at first a faint twilight, but gradually strengthening as the age advanced. At its close, "day and night—the one still dim and gray, the other wrapped in a pall of thickest darkness—would succeed each other as now, as the earth revolved on its axis, and the unseen luminary rose high over the cloud in the east, or sunk in the west beneath the undefined and murky horizon." So great was the heat that prevailed while the azoic strata were laid down that it does not seem strange that no remains of life are found in them. "As well," says Agassiz, "might we expect to find the remains of fish, or shells, or crabs at the bottom of geysers or of boiling springs, as on those early shores, bathed by an ocean of which the heat must have been so intense."

272. **State of the Surface at the End of this Age.**—At the end of the Azoic age there was nothing like the diversity of surface that there is at the present time, there being no high elevations over the comparatively small portions of land which then rose above the level of the universal ocean. Still, denudation had been going on during all the progress of the age, and there were therefore extensive areas where there were gravel, sand, and some material ground up so fine as to be mud. This broken and ground material was spread not only here and there over the land, but over the bottom of the seas as now. In short, the earth was prepared for the coming life of the next age. A part of this preparation was the reduction of the temperature to such a point as was consistent with the existence of plants and animals.

## CHAPTER XIV.

### AGE OF MOLLUSKS, OR SILURIAN AGE.

273. **Dawn of Life.**—The earth having been prepared during the Azoic age for vegetation, life, both vegetable and animal, was now introduced upon the scene. That its introduction occurred at that time is inferred from the fact that the beginning of the life-record is found in the rocks that were then formed, or, in other words, that no fossils, no remains of life, have been found in the rocks that were formed previous to that period. There

are, indeed, certain contents of rocks of the Azoic age, from which some have argued that there were both vegetables and animals in existence then, and that they were so acted upon by the mechanical and chemical agencies of that early period that we now find only what was produced from them. This supposition may possibly be correct. If so, we have an example of that foreshadowing that I spoke of in § 262, the scanty life of the one age in this case preceding the full introduction of life in the other. But, at whatever time life was introduced, it was done by a distinct exertion of creative power. It was no result of chemical or mechanical forces already existing, for dead matter has no disposition or tendency in itself to produce the seed of a vegetable or the germ of an animal. All vegetables and animals have a parentage, and the beginnings of the lines of succession were products of creative power. In other words, each species, either vegetable or animal, was, when first introduced upon the earth, a distinct creation. Though the influence of circumstances may produce varieties in any species, no species can be derived from any other species. There is a disposition in some to discard this view. They seem to dislike the idea of a present deity, and to desire the removal of creative power as far back in time as possible, as if in the dim distance they get rid of some of the actual force of the power.

274. **Rocks of this Age.**—The rocks are of considerable variety—hard sandstones, limestones, slates, shales, flagging-stones, marls, and conglomerates. Some of the formations are in part calcareous—that is, they have carbonate of lime mingled with other materials; while in some formations there is quite pure limestone, which in some localities has been converted into marble. At Niagara Falls we have 85 feet of limestone lying upon 80 of shales, the erosion of the soft shales by the water causing a constant undermining of the hard limestone, and therefore a recession of the falls toward Lake Erie, as stated in § 183.

**275. Arrangement of the Rocks.**—The arrangement of the rocks of this age was first thoroughly observed in a region of the western part of England, which was in ancient times inhabited by a tribe called the Silures. Hence this age is often called the Silurian age, and the rocks are said to belong to the Silurian system, no matter in what country they may be found. Local names have, after this fashion, been given quite extensively by geologists to different systems and formations, as you will see as we proceed. The Silurian system has been very fully examined in this country in the State of New York, and it is presented to you in Fig. 101, in a section which was drawn by Professor Hall in making his geological survey. The section exhibits the arrangement of the system, together with the next to be noticed, the Devonian, from the north side of Lake Ontario across New York into Pennsylvania. I give below it the names of the different formations or groups as they have been, applied from the local associations of the rocks.

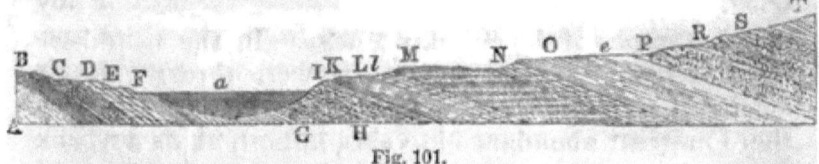

Fig. 101.

1. A. Upper rocks of the Azoic age.
2. B. Potsdam sandstone, lowest of the rocks of the Silurian age.
3. C. Calciferous sand rock.
4. D. Black River limestone.
5. E. Trenton limestone.
6. F. Utica slate.
   a. Lake Ontario.
7. G. Hudson River group.
8. H. Gray sandstone and Oneida conglomerate.
9. I. Medina sandstone.
10. K. Clinton group.
11. L*l*. Niagara group.
12. *m*. Onondaga salt group.
13. N. Helderberg series.
14. O. Hamilton group.
15. *e*. Tully limestone.
16. P. Portage group.
17. R. Chemung group.
18. S. Old red sandstone.
19. T. Conglomerate of the Carboniferous system.

You see that the rocks all crop out here, and that, if they laid horizontally, as they did when they were deposited, the azoic rocks, A, would be underneath the whole. The Silurian system ends in the Helderberg series, and the Devonian lies between this and the beginning of the Carboniferous system, T. The old red sandstone, S, is so prominent in the Devonian system in England and Scotland that its name is there very commonly applied to the whole system.

276. **Geographical Distribution.**—The chief part of the rocks of this system that come to the surface in the United States lie along south of the continent as it was begun in the Azoic age, as represented in § 267. They are in the great lake region, and extend along in a broad belt west from Lake Michigan. A long strip of this system stretches along on the east of the great Appalachian coal-field, that extends from Pennsylvania down into Alabama. The Silurian rocks are found in the west of England, where, as I have before said, they were first successfully examined; also in Belgium, Germany, Norway, Sweden, and in Russia, in the neighborhood of St. Petersburg.

277. **Copper of the Silurian Rocks.**—In the neighborhood of Lake Superior trap rocks were thrust upward among the Silurian rocks, and native copper is found there in great abundance in veins, in both kinds of rocks where they came together. It is supposed that the copper was produced from ores in the azoic rocks lying below, the trap, as it came up in its molten state, smelting the ores as if in a vast furnace, and forcing the pure metal upward, which, in its melted state, passed into openings and crevices, making veins in all directions.

278. **Silurian Salt.**—Most of the salt-springs in this country issue from Silurian rocks. The extensive accumulation of salt in the State of New York Professor Dana accounts for in the following manner. He supposes that when the saliferous (salt-bearing) beds were laid down there was in that quarter a large shallow basin, or series of basins, with limestone, such as we have

at Niagara Falls, for a bottom. Into these basins the salt water of the ocean was admitted, in some way, intermittingly. This might be from the occasional breaking away here and there of the barriers which bounded in these basins from the sea, or simply from the changes of the tide. The result, you see, then, would be somewhat like that which we have in the artificial production of salt from evaporation of the waters of the ocean, as noticed in § 358, Part II. By evaporation over this immense area of the basins the salt of the sea-water would be deposited, mingled with the mud, which would be at the same time settling as sediment. This would go on rapidly whenever the water was very low, especially with the warmth of climate which prevailed in that age. And such accumulations of salt going on for a long period (for this salt-making, like the coal-making, did occupy a long time in the world's geological history) would lay up in the rocks into which this mud would change by solidification a vast amount of salt, such as we now find in the Silurian deposits in New York. Mr. Dana states that he once saw in a small coral island of the Pacific a process similar to this actually going on. It was an island with a lagoon in it, having no free communication with the ocean. The waters became extremely salt in the hot, dry season, and fresh again in the rainy months. There was a mud deposited in the bottom of the lagoon from the washing of the coral rocks that bounded it. This calcareous mud, if solidified, would make a saliferous rock, like the rocks of this character in the Silurian system of New York. The salt is obtained from these rocks in New York by evaporating the solution, or brine, made by the water which gains access to the strata. At the great salt-works in Salina and Syracuse this brine is collected by borings, which sometimes extend down over 300 feet. From 35 to 45 gallons furnish a bushel of salt, while it requires nearly ten times this quantity of sea-water to furnish this amount.

279. **Gypsum.**—There is also much gypsum in the rocks of the saliferous epoch of the Silurian age in New York. It is found, not in layers, but in masses imbedded in the rocks, from some of which it is supposed to have been formed by a chemical process. The explanation of the process is this: The sulphureted hydrogen issuing from sulphur springs, which abound in New York, by becoming oxydized, produces sulphuric acid, and this, acting upon the limestone (carbonate of lime), unites with the lime, driving off, of course, the carbonic acid (§ 60). Wherever gypsum and salt are found together, as they often have been, the salt, being soluble, may be carried off by water, if circumstances allow of drainage, leaving the insoluble gypsum behind. In many cases where gypsum is found alone there was once salt in company with it.

280. **The Silurian Beach.**—As the land formed in the Azoic age was not elevated to any great height, the land which was added to it during the age of Mollusks was mostly added as accumulations are now made on a widespread beach by the waves of the sea. So much was this the case, that Professor Agassiz remarks that "in the Silurian period, the world, so far as it was raised above the ocean, was a beach." He speaks also of certain marks of water movements, usually seen on beaches, which are now found in the solid rocks of this age. Even ripple-marks made, not merely centuries, but thousands of centuries ago, are to be plainly seen now in the rocky strata as we divide the laminæ and expose fresh surfaces. So there are ridges left which are manifestly remains of ancient sea-shores, formed one after another as the land, with the accumulation of deposits, encroached upon the water. There are many of these ridges to be seen extending from the neighborhood of Lake Champlain toward the west, marking the limits of the sea at successive periods in the age. They have all the irregularities of line and shape that are witnessed in sea-

shores of the present day. "Unstable as water" is a proverb, the truth of which has been recognized from time immemorial; but the water, though itself unstable, has left stable evidences of its work every where in the earth. It has left traces in rock which have lasted ages upon ages longer than any chiselings of man upon monumental granite.

281. **Life in this Age.**—As the beach-arrangement of land was prominent, so, as Agassiz remarks, the life was mostly such as we find now on beaches. Mollusks were abundant; crustaceans of some kinds also, and star-fishes, sea-urchins, etc. There were corals, also, when and where the circumstances were favorable to their growth. I will cite two instances in contrast in this respect. In forming the limestone floor of the great New York basins preparatory to the salt-making (§ 278), the work was done by coral animals, and other allied animals, as is shown by the abundance of their remains in that rock. But when the salt-beds came to be laid down these animals were absent, for the water was too salt for them to live in, and few fossils are therefore found in these beds. The same contrast was seen in the coral island observed by Professor Dana. In the salt water of the lagoon there were no corals nor shells, though there was a plenty of them outside of the island in the open sea. There were no insects in the Silurian age, and but few fishes; some think absolutely none. There were no land nor fresh-water animals. As to vegetation, we have remains only of the lowest forms of vegetation, such as sea-weeds and club-mosses. No land-plants have been found except in the uppermost strata. What life there was in this age was very abundant during most of the period in most localities. There was such a profusion, says Agassiz, "that it would seem as if God, in the joy of creation, had compensated himself for a less variety of forms in the greater richness of the early types."

282. **Adaptation.**—The life on the earth at that time,

I

we can see from our reading of the life-record of the rocks, was adapted to the circumstances of the case. Though all the four grand divisions of animals were represented, there were few vertebrates or articulates, and these were of the lower orders, because the circumstances of the world were not suited to those forms of life. The animals were mostly marine. There were no air-breathing animals, because the atmosphere was not in a fit state for them. There was too much carbonic acid in it, of which the air was to be relieved in after ages, as will be shown you in another chapter. There were no fishes, partly, at least, because they need dissolved in the water which passes through their gills air that has a good proportion of oxygen in it (Part II., § 93). In regard to the adaptation manifest in that age of the earth, Agassiz very forcibly remarks: "Let us remember, then, that in the Silurian period, the world, so far as it was raised above the ocean, was a beach, and let us seek there for such creatures as God has made to live on sea-shores, and not belittle the creative work, or say that he first scattered the seeds of life in meagre or stinted measure, because we do not find air-breathing animals when there was no fitting atmosphere to feed their lungs, insects with no terrestrial plants to live upon, reptiles without marshes, birds without trees, cattle without grass—all things, in short, without the essential conditions for their existence."

283. **Carbonaceous Matter in Silurian Shales.**—There was no coal formed in this age, for there was no vegetable growth sufficient for this purpose; but in the shales there is a considerable amount of carbonaceous matter in some localities, even, in some cases, up to 20 per cent. This came mostly from the sea-weeds, which, gathering the carbonic acid from the air, appropriated to their own growth the carbon, and gave the oxygen to the animals of the sea (Part II., § 95).

284. **Hydrozoa and Bryozoa.**—In Fig. 102 we have the

AGE OF MOLLUSKS. 195

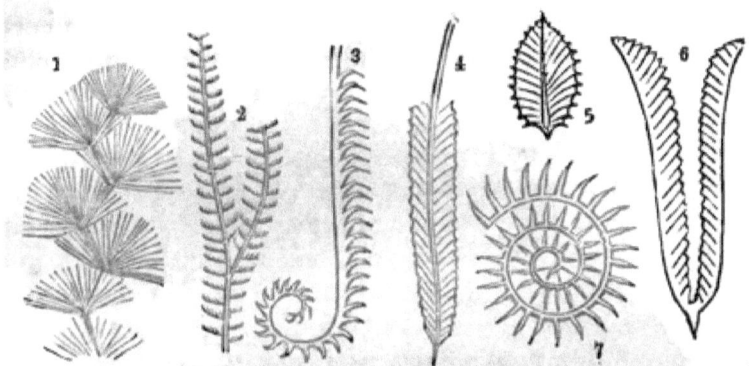

1. Oldhamia. 2. Protovirgularia. 3. Graptolites. 4, 5. Diplograpsus. 6. Didymograpsus. 7. Rastrites.
Fig. 102.

frame-works of certain small animals called Hydrozoa and Bryozoa, some of which contributed much to the material for the formation of rocks in the Silurian age.

285. **Corals and Echinoderms.**—In Fig. 103 we have specimens of these two divisions of the Radiates, which

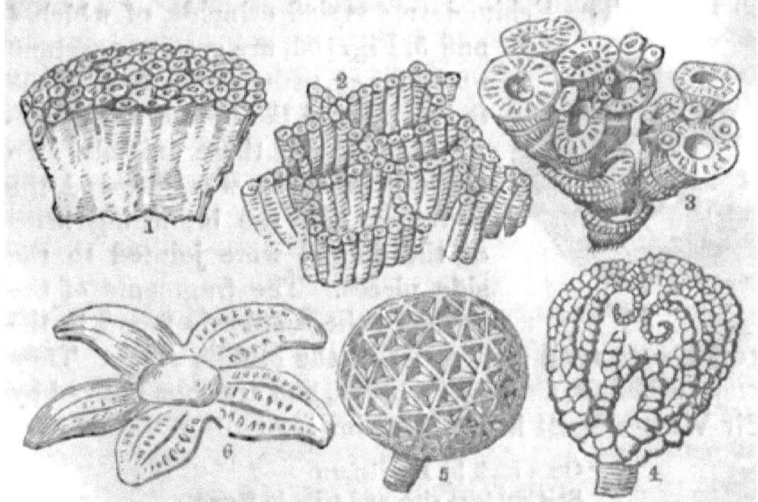

1. Heliolites. 2. Catenipora. 3. Cyathophyllum. 4. Taxocrinus. 5. Cystidea. 6. Palæaster.
Fig. 103.

had so much to do with supplying the material for the limestones in the age of Mollusks. One of the most beautiful of the corals is given in Fig. 104, the chain coral, the specimen pictured being from the cliff limestone

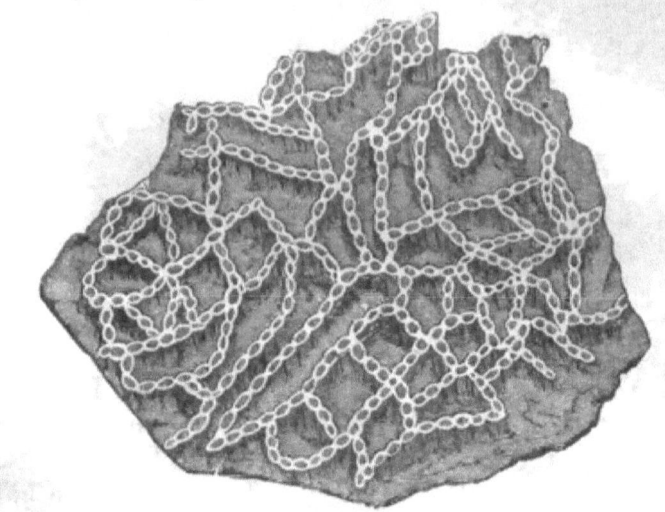

Fig. 104.

in Iowa. The Echinoderms styled crinoids, of which 4 and 5, Fig. 103, are specimens, stand on stems or peduncles. In Fig. 105 is represented the base of the framework of one of these animals. To the central piece was fastened the peduncle, and the branching arms of the animal were jointed to the side pieces. The fragments of the stems of these animals found in the rocks were used as rosaries in the Middle Ages. They were called St. Cuthbert's beads, and are thus noticed by Sir Walter Scott in his Marmion:

Fig. 105.

"On a rock by Lindisfarn
St. Cuthbert sits, and tries to frame
The sea-born beads that bear his name."

286. **Mollusks.**—In Fig. 106 you have specimens of the shells of some of the varieties of Silurian mollusks. There are three great classes of mollusks. 1. The *Cephalopods*,

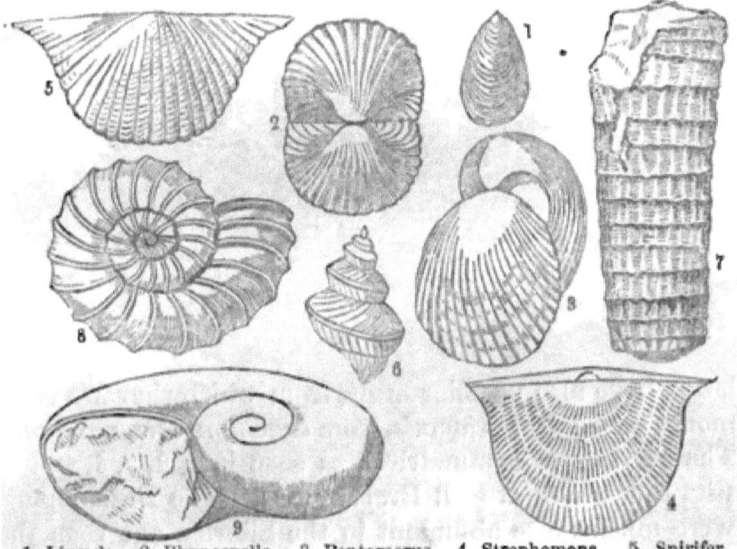

1. Lingula. 2. Rhynconella. 3. Pentamerus. 4. Strophomena. 5. Spirifer.
6. Murchisonia. 7. Orthoceras. 8. Lituites. 9. Maclurea.
Fig. 106.

which have arms or feet about the mouth, the name coming from the Greek works *kephale*, head, and *pous*, foot. The shells of this class are more often curved on one plane, as 8, than spiral, as 6. In 8 we have a *chambered* shell—an arrangement that is found in almost all of this class—the cavity being divided by partitions into chambers, which have a tube, called the siphuncle, running through them. The use of this is explained in my Natural History, § 554. 2. *Gasteropods.* The name of this class comes from *gaster*, belly, and *pous*. These mollusks creep on a broad disk or foot, as you see represented in Fig. 107 (p. 198). The head and the broad foot, which are out of the shell as the animal crawls along, can be withdrawn within the shell. The common snail is an example of this class. 3. The *Acephals.* These have no heads, as the name indicates, the *a* being the *a* privative of the Greek. Of these there are two divisions—the common bivalves, as oysters and clams, which have two equal valves, and those which have two unequal valves. These

Fig. 107.

latter have also peculiar arms, from which they are commonly called *Brachiopods*, from *brachion*, arm, and *pous*. The valves are symmetrical, as seen in 1, 2, 3, 4, and 5, Fig. 106, which are all Brachiopods. The Brachiopods were much more abundant in the Silurian age than the common bivalves, but the reverse is the case at the present time. The small shell, Lingula, 1, is sometimes found in rocks in such abundance as to give them their name, as Lingula flags and Lingula grits. The Orthoceras, 7, is a chambered shell, like the pearly Nautilus of our day, or the Lituites in the figure, but as if unrolled, so as to be straight. Some specimens are ten or fifteen feet in length and a foot in diameter. These were the monster mollusks of the Molluscan age. In Fig. 108 is represented the internal arrangement of the shells of these Orthoceratites. The Cephalopods, with their chambered cells, were very numerous in the Silurian and other ancient ages, but at the present time there are about half a dozen living species, and these belong to the genus Nautilus. In Fig. 109 we have a representation of a remarkable shell of a cephalopod from the Silurian rocks of Russia.

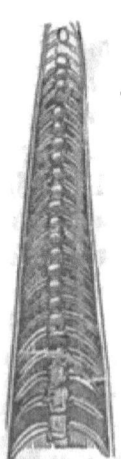

Fig. 108.

287. **Trilobites.**—The family of Trilobites, of which there is not a species now in existence, furnished the

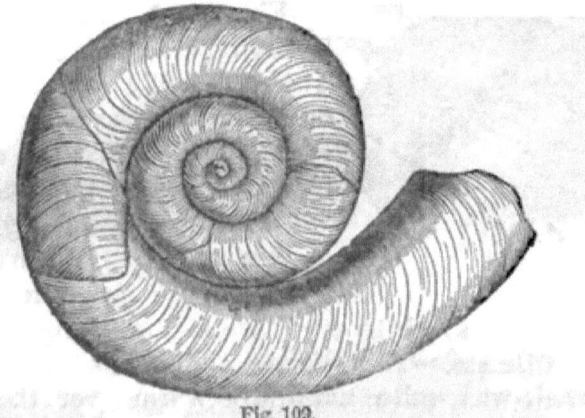

Fig. 109.

chief representatives of the division of Articulates called Crustaceans in the age of Mollusks. Some of the species of this extensive family are shown in Fig. 110. These

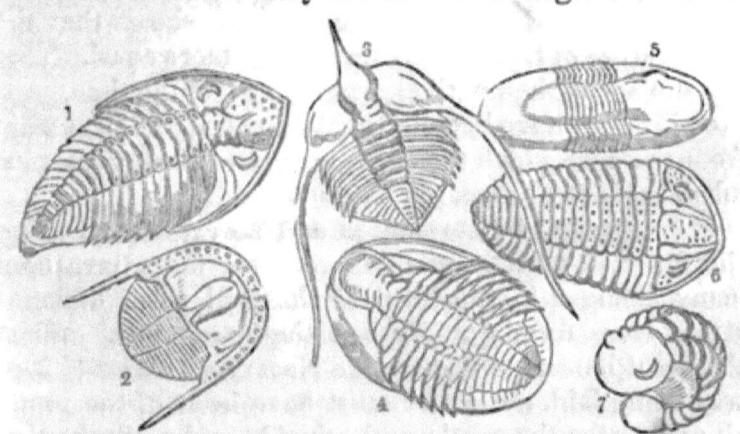

1. Phacops. 2. Trinucleus. 3. Ampyx. 4. Ogygia. 5. Ilænus. 6. Calymene. 7. Calymene coiled up.

Fig. 110.

animals, as you see, are divided lengthwise into three lobes, and hence comes their name. They have, transverse to these, the common ring-like divisions of the Articulates. Their eyes, which were quite prominent, were compound. They had the power of curling themselves up like the wood-louse, as exhibited in Fig. 111 (p. 200).

Fig. 111.

In Fig. 110, 7 is the same with 6, but represented as coiled up. The fossils of these animals are often found in this posture in the rocks, showing that the enveloping mud or sand, which afterward became stone, caught them in this condition. The size of these animals varied much, from the sixth of an inch to even two feet in length.

268. **Climate.**—The climate of the Silurian age, it is supposed, was quite uniformly warm over the whole earth. Mollusks are not apt to be profusely abundant except in warm seas, but, as the life-record shows that they were so in such localities as New York, it is plain that tropical climes were more extensive than now. Besides, the wide diffusion of the mollusks shows that the temperature of the earth generally was more equal. That the sun shone then with clearness much of the time, however it might have been in the Azoic age (§ 271), is plain from the eyes which the Creator gave to those numerous inhabitants of the sea, the Trilobites.

289. **Alternate Subsidences and Elevations.**—During the different periods of this age there must have been many changes in the level of the land when different strata were forming. Take a single example. When the great limestone floor of the New York basins (§ 278) was being laid, the water must have been of the requisite depth for the corals, and crinoids, and mollusks that flourished then. For this purpose there was a subsidence of the land, and there is evidence that this movement extended far beyond the locality of these basins. For instance, the Green Mountain region, which had previously been dry land, was now wholly or in part submerged. But when, after the floor was laid, the salty deposits were made, there was an elevation of the land that shallow waters might prevail for the ready evaporation of the water and the consequent deposit of the salt.

## CHAPTER XV.

### AGE OF FISHES, OR DEVONIAN AGE.

290. **Rocks.**—The rocks formed in this age are, in North America, in the first part of the age, limestones, and afterward mostly sandstones, shales, and flags. The famous North River flagging-stones were made in this age. So also were the Scotch flagging-stones known over Europe as the Caithness flags of commerce. The Devonian system of England and Scotland has the general name of Old Red Sandstone, red sandstone being there the principal rock of the system. The name Devonian was given to this system because its rocks were early investigated by Murchison in Devonshire, England.

291. **Extent of the System in this Country.**—The Devonian rocks cover a large area in the interior of this country, extending along south of the addition which was made in the age of Mollusks to the continent as it was begun in the Azoic age. Lake Erie is included in this space. Lake Michigan bounds it on the west, and a long, narrow arm of it extends from a little below this lake far to the northwest. This area extends broadly down the west side of the Appalachian coal formation two thirds of its length, tapering, however, as it goes. This immense Devonian field has a broad eastern branch extending from Lake Erie to within a short distance of the Hudson River. It is this part of the field, reaching down into Pennsylvania, which is represented in the section given in Fig. 101, beginning in N, the Helderberg series, and ending in S, the Old Red Sandstone. Perhaps some of this large extent of territory has had some of the Devonian strata covered by other rocks of a later

formation, but, if so, they have been denuded by the action of water.

**292. Devonian System in Europe.**—Devonian rocks are found in most of the countries of Europe. In Germany they cover a larger space than the Silurian. This system is very marked in England, and also in Scotland, where the genius of the lamented Hugh Miller has given it a deep but melancholy interest. In Russia, south of St. Petersburg, it covers an area of 150,000 square miles. In connection with the Devonian of Russia, I will mention an interesting example of geological investigation. It was doubted for some time whether the red sandstone of Devonshire was of the same period with that of Hereford, because the strata of Devon contained certain shells which were not found in the rocks of Hereford, while the latter contained fishes that were not found in those of Devon. Sir R. Murchison settled this question by finding in Russian strata both the shells of Devon and the fishes of Hereford. You see what this shows—that the rocks in Russia, in Devon, and in Hereford belong to the same epoch, and that the fact that the fishes and shells were not found together in both Devon and Hereford was owing to some local causes.

**293. Coral Formations.**—I have stated in § 290 that the strata of this age first laid down were limestones. The corals had a great agency in providing material for these. Indeed, the upper Helderberg period was, Professor Dana states, "eminently the coral-reef period of the Palæozoic ages." Near Louisville, Kentucky, at the Falls of the Ohio, there is a grand display of the limestone of this period, very much like a coral reef of the present time. At all seasons when the water is not high a series of ledges is exposed, and the softer parts of the rock having been worn away, the harder corals stand out in bold relief, and many of them branch out precisely as if they were living. There are honey-comb, cup-shaped, star-like, and other forms mingled with the joints, stems, and heads of crinoids. Among them is seen a species of the family Favosites, having a beautiful honey-comb structure, and appearing in masses in some cases of not less than five feet in diameter. A species of this family is represented in Fig. 112. The elegant chain coral, Fig. 104, is also present in this locality.

Fig. 112.

294. **Variety in Rock-making at the same Period.**—Different kinds of rock-making were sometimes going on at the same time over different parts of the Devonian area in this country, and the same is true of other countries also. Thus, at one period, while corals, crinoids, etc., were accumulating limestones over a large area in the Western States, over New York there were extensive flats forming and solidifying into sandstones, and shales, and flags. Here are found ripple-marks and shrinkage cracks, showing that the flats were sometimes covered with shallow water, and sometimes drying in the sun. The ripple-marks in the flags are so even and extensive that it is obvious that the sea swept over widespread flats. While this was going on in New York, there was at the West, where the corals and crinoids were building limestone, an interior sea of vast extent, and of suitable depth for these animals to flourish in.

295. **Devonian Vegetation.**—In the age of Fishes there was not only more marine vegetation than in the age of Mollusks, but land vegetation appeared, and before the middle of the age had passed it had covered the comparatively small portions of the continents that had fairly risen above the ocean and remained so. At the close of the age the State of New York, although it had been previously submerged, formed a part of the land which was thus covered. The plants, as we approach the conclusion of this age, become more and more like those of the next age, thus exemplifying that foreshadowing of

204  GEOLOGY.

which I have spoken in § 262. Indeed, there has actually been found some coal in some of the rocks, in addition to the carbonaceous shales, which appear now as they did in the Silurian system.

296. **Animal Life.**—As there was a considerable advance upon the previous age in the forms of vegetable life, there was a corresponding one in those of animal life. Few, if any, vertebrates existed in the Silurian age, but now they abounded in the form of fishes. Some reptiles also appeared, though not in much variety. There is no evidence of the existence of any insects, or birds, or mammals.

297. **Mollusks.**—Of the Mollusks there was as abundant a variety of species as in the previous age. Of the Bra-

Fig. 113.

chiopods, the genus Strigocephalus, of which Fig. 113 represents one species, was introduced at this period. Of the Cephalopods, the genus Clymenia was introduced, of which Fig. 114 represents one species. Thirty-five species, which is nearly

Fig. 114.

all under this genus, have been found in the rocks of one locality in Bavaria.

298. **Crustaceans.**—The Trilobites, whose species were numbered by the hundreds in the Silurian age, now amounted to a dozen or two. But there are some very extraordinary Crustaceans, which, beginning to appear in the age of Mollusks, were now very abundant. They

were singular in their formation, combining in one animal the qualities of various kinds of animals. "King-crab-like," says Page, "in their carapace and organs of mastication, lobster-like in their prolonged and segmented bodies, furnished with broad, paddle-like swimming limbs, and frequently with huge prehensile claws, they present the zoologist with an entirely distinct family, if not with the elements of a new and separate legion. Some of the species are of great size—three, four, and six feet in length—and seem to have been the scavengers of their period, living on the lower forms and garbage of the sea-shore." One of these is pictured in Fig. 115, lying on his back. In the same rocks in which are

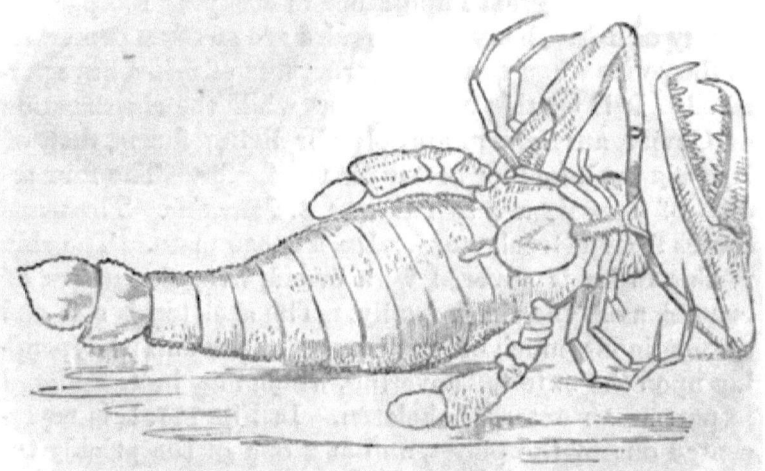

Fig. 115.

the remains of these Crustaceans there are found patches of spawn-like appearance, which, though they have been supposed by some to be berries of some plant, are now generally regarded as the egg-packets of these animals. The eggs are flattened, and they have more or less of a concentric arrangement, as seen in Fig. 116 (p. 206).

299. **Fishes.**—That you may understand the record which these animals have made of themselves in the rocks of this and other ages also, I will here notice the

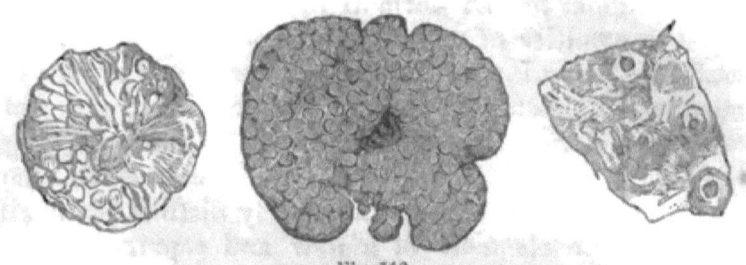

Fig. 116.

classification of fishes. Agassiz discovered that there is such a relation between the form of the scale and the internal organization of the fish, that those which have similar scales are similar in their general character. This discovery is of great importance in studying the palæontology of fishes, because the scales are so often preserved in the rocks when the other structures of fishes have perished. It is on this account that while the classification of Cuvier answers admirably for living forms, that of Agassiz is altogether better for the fossils. The four orders of Agassiz are as follows: 1. *Placoids*. The name comes from a Greek word, *plax*, a broad plate. The skin in this order is covered with broad, irregular plates of enamel, as in the shark family. The skeleton is soft and cartilaginous, much of the firmness of the animal depending upon the external covering, which may be considered in part as an external skeleton. In Fig. 117, 1, is represented one of the plates, and at 2 one of the prickly tubercles of the ray-fishes, which belong to the same order. 2. *Ganoid*. This term comes from *ganos*, splendor. Fishes of this order are covered in a regular manner with scales of horn or bone, having a thick outer layer of enamel, which is hard and bright. Such a scale is figured at 3. The sturgeon and trunk-fish belong to this order. 3. *Ctenoid*. This name, from *kteis*, a comb, indicates scales, 4 and 5, which are toothed or jagged on the posterior margin. The perch is an example of this order. 4. *Cycloid*. The term comes from *kuklos*, a cir-

AGE OF FISHES. 207

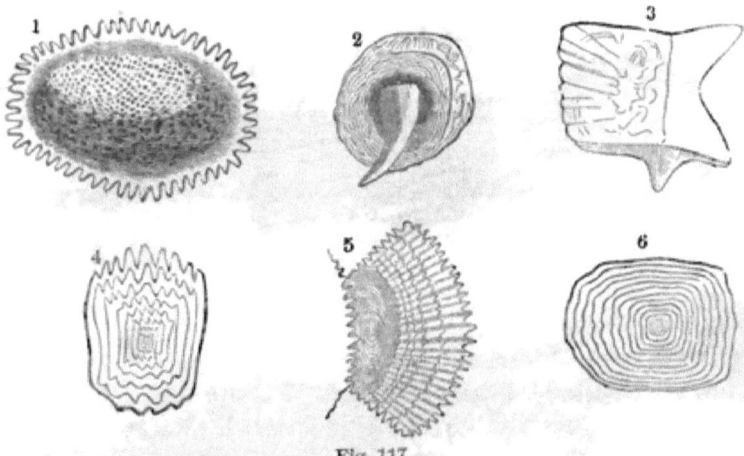

Fig. 117.

cle. The scales, 6, have a simple and smooth margin, and their outer surface is often variously ornamented. In this order are the salmon, the carp, and the pike.

I shall have occasion to refer to these orders hereafter. Suffice it to say now that the placoids and ganoids had their greatest developments in the Palæozoic ages, and that there are few of them at the present time, while the ctenoids and cycloids came on at a later age, and abound in the waters of the present period. Four fifths of the present fishes belong to the ctenoid and cycloid orders, the remaining fifth consisting of placoids, with a small number of ganoids.

300. **Coccosteus, Pterichthys, and Cephalaspis.** — The fishes of the Devonian age whose names I have here given are ganoids, having very peculiar characteristics. The Coccosteus, or "berry-bone," 1, Fig. 118 (p. 208), has the largest part of the body incased in a box-like covering of bony plates, upon which there are berry-like projections. The Pterichthys, or "wing-fish," 2, which Hugh Miller, its discoverer, termed the characteristic organism of the old red sandstone, is so extraordinary that Agassiz says of it that "it is impossible to find any thing more eccentric in the whole creation." On the head was a

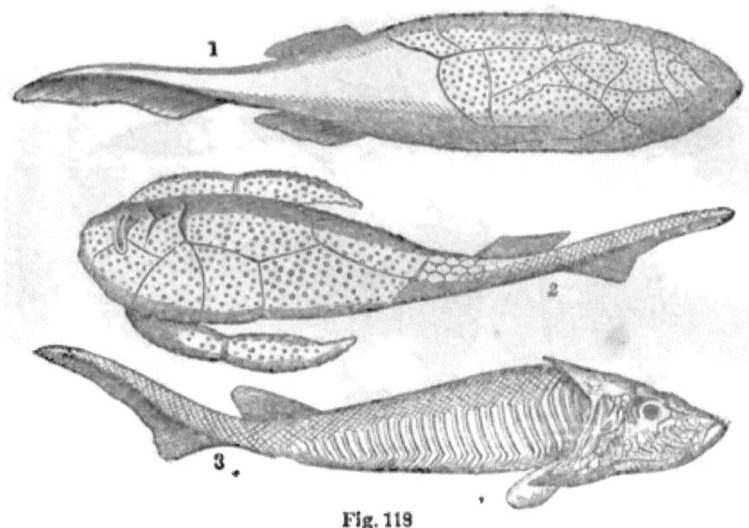

Fig. 119

strong helmet with two circular holes in front for the eyes. The chest and back were covered by a curiously-constructed cuirass formed of plates, and the tail was sheathed in a flexible mail of bony scales. The arms are also covered with plates. There are peculiarities in the jointing of the different parts of this bony covering, and in the contrivances for securing lightness with strength, which are very curious. Hugh Miller says of it that, "with its inflexible cuirass and its flexible tail, and with its two arms, that combined the broad blade of the paddle with the sharp point of the spear, it might be regarded, when in motion, as a little subaqueous boat mounted on two oars and a scull." He farther says that "when, in laying open the rock in which it lies, the under part is presented, as usually happens, we are struck with its resemblance to a human figure, with the arms expanded, as in the act of swimming, and the legs transformed, as in the ordinary figures of the mermaid, into a tapering tail." The correctness of this description can be seen in Fig. 119. The Cephalaspis, or "buckler-head," 3, Fig. 118, is so called from the shield-like shape of the bony head-plate, which is all in one piece.

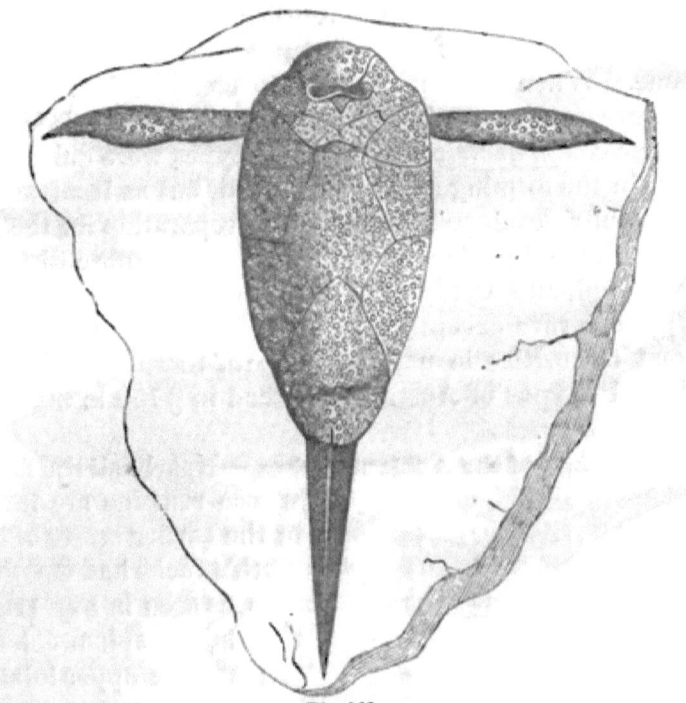

Fig. 119.

301. **Asterolepis.**—This is another of the ganoids of the Old Red Sandstone. It equaled in size the largest of alligators. The helmet on its head was made of strong bony plates, ornamented with star-like markings, and its body was covered with a mail of scales which looked as if they had been carved in the most delicate manner. It is stated by Hugh Miller that helmets of this creature "have been found in the flag-stones of Caithness large enough to cover the front skull of an elephant, and strong enough to have sent back a musket bullet as if from a stone wall." But what was most strange of all is that, although the Asterolepis was a fish, it had in the arrangements of its jaws and teeth the complete characteristics of a reptile. Here is an example, in another form, of the foreshadowing spoken of in § 262. We see, in this age of Fishes, stamped unmistakably upon some of these an-

imals characteristics of the reptiles which are to abound so largely in one of the following ages as to give it its name. When such mingling of the characteristics of two or more classes of animals occurs in one animal, it is said to be a *comprehensive* type. Such types were quite common in the forming ages of the world, but as it advanced to its fully-developed condition, in preparation for the advent of man, the classes of animals became more distinctly defined, and at the present time such comprehensive types are rare exceptions—so rare as to be regarded as great curiosities in nature. One of these is the duck-billed Platypus of Australia, noticed in § 133 in my Natural History.

302. **Tails of the Ancient Fishes.**—Nearly all the fishes whose remains are found in the earlier strata of the earth's rocks had unilobed tails, as seen in Fig. 120— that is, the spinal column ran into the upper lobe of the tail, the lower being so small as not to be counted as a lobe. In the later strata, on the other hand, the tails are bilobed, Fig. 121, in the great majority of the fossils. At the present time the only family of fishes that has the unilobed tails is the Shark family.

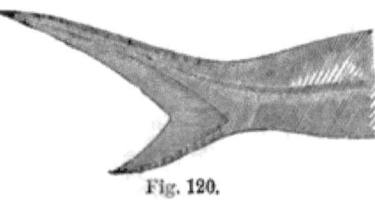
Fig. 120.

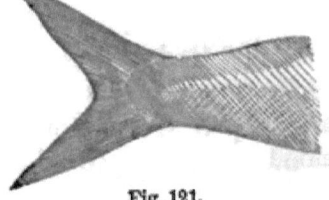
Fig. 121.

303. **Abundance of Devonian Fishes.**—In estimating the abundance of ancient fishes from their fossils, we are to consider how uncommon a thing it must be for fishes to become enveloped in sand or mud, in comparison, for example, with shells. Keeping this in view, the great number of species that have been made out from the fossils in some quarters, especially in Great Britain and Europe, indicates that fishes must have been very abundant in the Devonian age. In some cases great numbers must,

from some cause, have been suddenly destroyed and buried at the same time, for Hugh Miller states that there are strata in Scotland where they lie as thickly as herrings do on the fishing-banks in the time of the fisherman's harvest. And often there is positive proof of their having been suddenly killed and buried in their contorted attitudes. Undoubtedly, in such cases, the fishes were overwhelmed with the material in which they were enveloped by some great convulsion, and then they became fossilized in the rock into which this material was converted by solidification.

304. **United States at the End of this Age.**—I have given you in § 267 some idea of the beginning of the continent of North America in the Azoic age, and in § 276 some idea of the additions which were made in the Silurian age. During the Devonian age additions were made on the south and southwest, carrying out the plan already begun, as indicated in Fig. 100. Agassiz has stated what these additions were. He says that so much was added that at the conclusion of the age there were above the water within the United States "the greater part of New England, the whole of New York, a narrow strip along the north of Ohio, a great part of Indiana and Illinois, and nearly the whole of Michigan and Wisconsin." Besides this, from upheavals at the close of this period, there were, in place of the Rocky Mountains at the West, and of the Appalachian chain, which now stretches as a rocky wall from New England to Alabama, detached islands and reefs amid shallow waters. One of the upheavals that occurred at the conclusion of the Devonian age raised the high ground on which the city of Cincinnati now stands. Of this hill Agassiz says, "The granite did not break through, though the force of the upheaval was such as to rend asunder the Devonian deposits, for we find them lying torn and broken about the base of the hill; while the Silurian beds, which should underlie them in their natural

position, from its centre and summit. This accounts for the great profusion of Silurian organic remains in that neighborhood. Indeed, there is no locality which forces upon the observer more strongly the profusion and richness of the early creation, for one may actually collect the remains of Silurian shells and Crustacea by cart-loads around the city of Cincinnati. A naturalist would find it difficult to gather along any modern sea-shore, even on tropical coasts, where marine life is more abundant than elsewhere, so rich a harvest, in the same time, as he will bring home from an hour's ramble in the environs of that city."

305. **Scenery of this Age.**—During most of the Devonian age the scenery must have been exceedingly tame and monotonous. "Over dark and shallow seas," says Hugh Miller, "mud-banks of vast extent occasionally raised their flat, dingy backs, and remained hardening in the hot sun until their oozy surfaces had cracked and warped, and become hard as the sun-baked brick of Eastern countries; and then, ere the seeds of terrestrial plants, floated from some distant island, or wafted in the air, had found time to strike root into the crevices of the soil, some of the frequent earth-tremors of the age shook the flat expanse under the water out of which it had arisen, and the waves rippled over it as before." And when vegetation obtained any where a foothold upon the land, as was the case quite extensively in the latter part of the age (§ 295), it was comparatively scanty, and there was none of that variety of surface which we have now, for there were no lofty mountains nor large rivers. Besides, there were no birds nor four-footed beasts to enliven the scene.

## CHAPTER XVI.

### THE CARBONIFEROUS AGE, OR AGE OF COAL.

**306. Propriety of the Name.**—Nothing has been made more plain by the investigations of geologists than that thousands of centuries ago there was an age, and that a very long one, devoted expressly by the Creator to the formation and storing up of coal for the future use of man. It seems eminently proper, therefore, to call this the Age of Coal, or the Carboniferous Age. Observe that all the other ages that transpired during the formation of the earth were marked by the predominance of certain classes of animals, and are named accordingly; but this is named from a mineral production which came, as you have seen in § 42, from certain chemical changes in vegetable substances. Animals of most of the various classes were abundant in this age, but no class was especially predominant over all the others, so that it is impossible to derive a name from that source. There seems to be no room for choice, then, but we are driven to the name which has been adopted by universal consent.

**307. Localities of Coal.**—There are more and larger beds of coal in North America than in any other part of the world. Great Britain comes next in order. The coal there is mostly bituminous, more or less, while in this country there are large quantities of the anthracite, or non-bituminous coal. Pennsylvania has a larger area of coal in proportion to its whole area than any other part of the world. There is coal in Spain, France, Belgium, and Germany. Russia exhibits over a wide area some of the rocks of the coal period, but has very little coal.

**308. Amount of Coal.**—The amount of coal in the principal coal-fields in the world has been estimated by Professor H. D. Rogers as follows:

| | |
|---|---|
| Belgium............................................. | 36,000,000,000 tons. |
| France.............................................. | 59,000,000,000 " |
| British Isles....................................... | 190,000,000,000 " |
| Pennsylvania..................................... | 316,000,000,000 " |
| Appalachian coal-field................... | 1,387,000,000,000 " |
| Indiana, Illinois, and Kentucky...... | 1,277,000,000,000 " |
| Iowa, Missouri, and Arkansas........ | 739,000,000,000 " |
| Total amount in North America.... | 4,000,000,000,000 " |

The anthracite of this country was first introduced into use in blacksmithing by Judge Obadiah Gore, a Connecticut blacksmith, in Wilkesbarre, a few years less than a century ago; but it did not come into common use until between thirty and forty years since. In 1820 the amount worked in Pennsylvania was only 380 tons. Its use, however, from that time increased rapidly, and in 1847 that state furnished over three million of tons, besides two million of bituminous coal. The United States produced in 1857 ten and a half million of tons; and if our annual consumption should be twelve million of tons, it is calculated that there is coal enough in North America to last us 333,333 years.

**309. Extent of Coal-fields.**—Coal-fields vary much in extent, the largest in the world being in North America. The most extensive of all is the Appalachian. This occupies parts of Pennsylvania, Ohio, and Virginia, the eastern portion of Kentucky and Tennessee, and extends down into Alabama. It covers an area of 80,000 square miles, 60,000 of this being available. This is about ten times as great a space as that occupied by all the productive coal-fields of Great Britain. Then there is the Indiana, Illinois, and Kentucky coal-field, covering a space of 50,000 square miles, and the Iowa and Missouri coal-field, 60,000. The New England coal-field occupies an area of only about 600 square miles.

**310. Arrangement of the Coal-strata.**—The coal is in

strata or seams of various degrees of thickness, between strata of different kinds of rocks. Sometimes the layers are almost as thin as paper, and, on the other hand, they are sometimes 30, 40, and even 60 feet thick. Seldom, however, do they exceed eight feet in thickness. The lower part, or floor, as we may term it, of the coal formation is limestone; but the upper part, where the beds of coal lie between strata of rock, is made up of sandstones, conglomerates, shales, and limestones. It is this upper portion that is commonly called the coal-measures. The rocks between which the coal is laid down do not differ from many rocks in other formations, but are to be distinguished from them only by their fossils. It is their life-record that indicates their proximity to coal. This fact is of great practical use in searching for localities of coal, and a disregard of it has sometimes occasioned needless expense. On this point Professor Hitchcock remarks: "No geologist would expect to find valuable beds of coal in the oldest crystalline rocks, but in the fossiliferous rocks alone; and even here he would have but feeble expectations in any rock except the coal formation. What a vast amount of unnecessary expense and labor would have been avoided had men who had searched for coal been always acquainted with this principle, and able to distinguish the different rocks! Perpendicular strata of mica and talcose schists would then never have been bored into at great expense in search of coal; nor would black tourmalin have been mistaken for coal, as it has been." Commonly, the rock that lies directly under a bed of coal is clayey in its character, and is called the *under-clay*. From its composition, and from the fact that roots of certain carboniferous plants are often found in it, it is inferred that it was the dirt-bed in which the plants grew that furnished the material for the formation of the coal. The amount of rock in the strata of the coal-measures preponderates vastly over the amount of coal, the proportion being generally fifty or more feet of rock to one of coal.

**311. Sub-carboniferous Period.**—That first part of the Carboniferous age in which the limestone floor was laid down for the coal-measures to rest upon is called by geologists the sub-carboniferous period. The prefix *sub* being the Latin for under, you see the propriety of the term. This floor was laid down over a very large area in this country. The same thing is found in other countries where coal was made in this age. The thickness of this formation varies much in different localities, in some amounting to more than 6000 feet. As this limestone was made from the remains of marine animals, crinoids, corals, etc., long ages must have been occupied in constructing this foundation of the coral-fields. The strata of the proper carboniferous rocks, the rocks of the coal-measures, are conformable with the sub-carboniferous.

**312. Sub-carboniferous Period in this Country.**—The circumstances under which the sub-carboniferous limestone was formed in the United States have been admirably brought out by Professor Dana, and I will attempt briefly to make the principal of them clear to you. In the Niagara period of the Silurian age there was a formation of limestone over a large portion of the interior of the United States. This was made by the deposition of the remains of marine animals. While this was going on all the interior part of the country was a sea, not very deep, but of sufficient depth for these animals. This sea might be considered as an extension upward of the present Gulf of Mexico. It was cut off from direct communication with the Atlantic Ocean by a low barrier of land stretching from the north downward, which had been made to emerge in order to shut in this great interior sea. A similar condition of things existed in the Devonian age, when, in the Upper Helderberg period, the corals were so busy in laying down limestone (§ 293). A similar arrangement existed also in the sub-carboniferous period, except that the interior sea was smaller than in the Devonian, as it was smaller then than in the Silurian. It grew smaller from the additions which were made in successive periods to the permanent land of the continent. There was another point of difference—in the sub-carboniferous sea the crinoids were predominant, so that Dana speaks of it as the Crinoidal sea, while in the Devonian it was the corals that did most of the work, and in the Silurian, brachiopods, corals, and crinoids were about equally present.

**313. How Coal was Made.**—All the coal in the world has been formed from trees and other plants. Proof of this has been adduced in § 41, and need not be dwelt upon now. The coal is the result of a decomposition, to a greater or less degree, of the woody fibre or substance of the wood. When wood is burned in the open air all the carbon is dissipated by gaseous combinations; but when the combustion is effected with the air mostly excluded, the carbon is retained to a considerable extent, making charcoal. For a full explanation of this I refer you to Part II. The same changes that occur in the combustion that produces charcoal may occur, under certain circumstances, without the phenomenon of combustion. If, for example, the decomposition take place slowly under water, the chemical changes are essentially the same as when what is ordinarily called combustion is seen. If you observe what the composition of wood is, you can see how this can be. Wood is composed of carbon, hydrogen, and oxygen, the elements which are engaged in ordinary combustion; and it is by the slow action of these upon each other that the slow combustion under water goes on. Very properly, then, is this process called *eremacausis*, this term being made from two Greek words, *erema*, slowly, and *kausis*, burning. The same thing occurs in the decay of vegetable matter in the open air, the slow combustion in this case being, however, carried out in full, as in the ordinary burning of wood, the carbon, oxygen, and hydrogen all forming

K

gradually new combinations, as they do quickly in common combustion. Heat is more or less an agent in effecting eremacausis, as it is in effecting the combustion which is sudden and attended with palpable phenomena.

314. **Condensation of Coal by Pressure.**—While the coal is forming it is subjected more or less to pressure by the masses of matter which accumulate above it, and form the rocky strata. It is therefore condensed, instead of being light and porous, like charcoal, or the coke which is obtained in the making of gas. In peat we have an example of the looseness of coal formation when subjected to but little pressure.

315. **Bituminous and Non-bituminous Coal.**—Anthracite, or non-bituminous coal, is almost pure carbon, but bituminous coal contains considerable hydrogen and oxygen. It is the hydrogen, in connection with the carbon, that produces its flame when it is burned. The flame is really the burning of the carbureted hydrogen, which in our gas-works is evolved from bituminous coal by a low degree of combustion, and is then passed about by pipes to be burned in our houses. The flame which plays over an anthracite fire when it is not wholly kindled is the burning of a different gas, which I have noticed in Part II., § 67. It is supposed that anthracite coal was once bituminous, and that it was freed from its bituminous qualities by a process essentially the same with that by which we make gas in our gas-works, the difference being simply that we have left in the one case a solid non-bituminous coal, and in the other a porous one—common coke.

316. **How the Coal was Deposited.**—You are now prepared to see how the coal was deposited in beds or strata between strata of rocks. In the first place there was a growth of plants of various kinds, varying in size from small mosses to enormous trees, in the dirt-bed of which I have spoken in § 310. This was, for the most part, a swampy growth, and that, too, as you will soon

see, of a most luxuriant and profuse character. Trees and shrubs dropped their leaves and fruit year after year, and at length died themselves, while other trees and shrubs took their places; and thus growth and decay went on, perhaps, through many slowly-moving centuries, till enough of these vegetable remains was accumulated to make a coal-bed. At length a subsidence of the land took place, the water flowed over it, covering up all this accumulation of vegetable substance, and under the water and the detritus which the water brought in, the decomposition occurred which resulted in the production of coal. The detritus went on accumulating, at the same time slowly solidifying from below upward, till at length the subsidence ceased, and there succeeded an elevation, so as to bring the surface again into the swampy condition for a new growth, preparatory to another bed of coal. The set of processes thus described was in some cases many times repeated, there being often many beds of coal between the strata of rock. In Kentucky there are from fifteen to twenty separate coal-beds in the strata, and in Nova Scotia, at a locality called the Joggins, there are seventy-six coal-seams, some of them being very thin. The process was essentially the same for each of these, the periods of elevation and subsidence being longer for the thick seams than for the thin ones. Though there were alternate subsidence and elevation, " the sinking condition," remarks Hugh Miller, " was the general one; platform after platform disappeared, as century after century rolled away, impressing upon them their character as they passed; and so the coal-measures, where deepest and most extensive, consist, from bottom to top, of these buried platforms, ranged like the sheets of a work in the course of printing, that, after being stamped by the pressman, are then placed horizontally over one another in a pile."

317. **Impurities of Coal.**—Coal varies much in its purity, as every one who has burned coal must have no-

ticed from year to year. The impurities came in part from the woody substance from which the coal was made. There was silex, which is one of the chief of the impurities, in some of the plants of the Coal age. But impurities came chiefly from the detritus which was laid down upon the coal-beds, for there would, of course, be more or less admixture of this with the coal. There is more or less iron pyrites (sulphuret of iron) in coal, and to this is owing the sulphur smell which a coal fire gives out. When there is much of it in coal it detracts much from its value, causing it to crumble on exposure to the air, and to emit strong fumes when burned.

318. **Rate of Formation of Coal.**—Some calculations have been made in regard to the time required for the formation of beds of coal. Taking some observations of Liebig as to the rapidity of vegetable accumulation, it has been calculated that it would require 170 years to make one inch thickness of anthracite coal, and therefore 122,400 years to accumulate a stratum of 60 feet. This is based upon the ordinary growth of the present time; but, as you will soon see, the growth of the Carboniferous age was probably much more rapid than it is now, and this, of course, must reduce the time. But, at any rate, the time must have been very long for the accumulation of even a bed of ordinary thickness. And when we come to take into the account the formation of the rocky strata as well as the coal-beds, our arithmetic must give out, for the whole system, from the sub-carboniferous upward, through all the strata, has a thickness, in some cases, of from 12,000 to about 15,000 feet—that is, nearly three miles.

319. **Plants.**—The remains of plants are found in great abundance in the coal-measures, though none appear in the great sub-carboniferous limestone floor which is beneath them. The plants were of various sizes. Many of them were prodigious in size, compared with similar kinds of plants at the present day. The vegetation of

that age was to a large extent, in point of size, a forest vegetation, and yet but few of the plants which at the present time are of the same family with the largest that existed then are a little taller than a man. The remains found are of various kinds—trunks, branches, pieces of bark, cones, leaves, etc. The impressions of leaves upon the laminæ of rock are sometimes very beautiful, as seen in Fig. 122. Here we have a carbonaceous representation of the stem and leaves—that is, the carbon of the plant in the decomposition of the vegetable substance is left upon the stone. "The shaly beds," says Dana, "often contain the ancient ferns spread out between the layers with all the perfection they would have in a herbarium, and so abundantly that, however thin the shale be split, it opens to view new impressions of plants." The experiments of Professor Göppert, of Breslau, go far to explain the various conditions in which we find the remains of plants in the rocks of the coal-measures. He placed fern-leaves in clay, and when they had become dry exposed them to a red heat, and thus obtained striking resemblances to the fossil ferns in the rocks. According to the degree of heat, the leaves were found to be either brown, shining, black, or entirely lost, the impression alone remaining. In this latter case the carbon of the leaves, being diffused in the clay, stained it black, thus showing that the color of the coal-shales is derived from the carbon of the plants inclosed in them. I will now proceed to notice some of the plants of this age.

Fig. 122.

320. **Calamites.**—This is a name given to a family of plants that belong to the same tribe with the horse-tails, cat-tails, and rushes of the present day. Two species

are represented in Fig. 123, *b* being the same with *a*, but

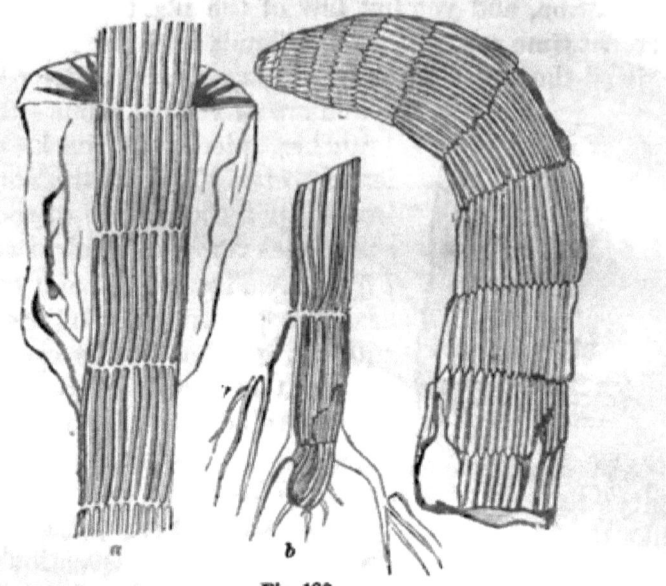

Fig. 123.

showing remains of its roots. The plant is jointed and hollow, the surface being deeply fluted. While the rushes and cat-tails of the present day are quite small, some specimens of these ancient plants have been found 20, and even 40 feet in length, and three feet in diameter. Portions of the trunks of these trees, for so they may be called, have been found standing upright in coal mines, penetrating the sandstone layers above the coal.

321. **Sigillaria and Stigmaria.**—These were for a long time supposed to be two distinct plants, but they have been found to belong to the same plant, the Sigillaria being the trunk, and the Stigmaria the lower part of the trunk, with the roots. The Stigmaria, Fig. 124, is never found any where but in the under clay spoken of in § 310. The trunks of the plant are generally found lying horizontally, but sometimes they are erect. They vary in length from five to sixty feet, and in diameter from a few inches to five feet.

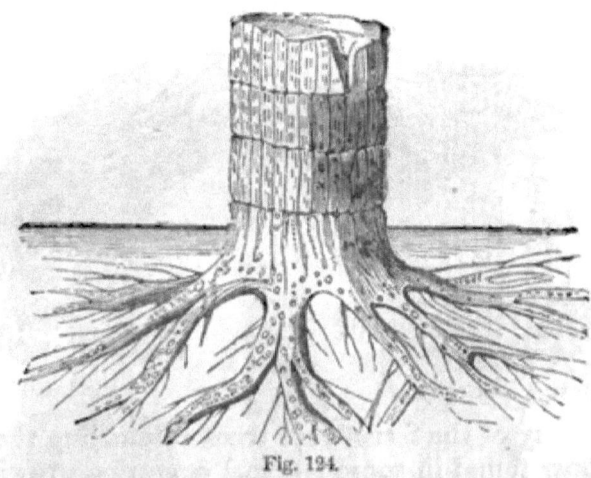

Fig. 124.

322. **Lepidodendron.**—This is a name which has been given to a tribe of carboniferous plants that were much like the club-mosses of the present day; but they were lofty, woody trees. One was found in a coal mine in England that was 40 feet long, and so enormously large at the base as to measure 13 feet in diameter.

323. **Ferns.**—There was a great abundance of ferns in the Carboniferous age. A very large number of species have been found in the coal strata. In Figs. 125, 126,

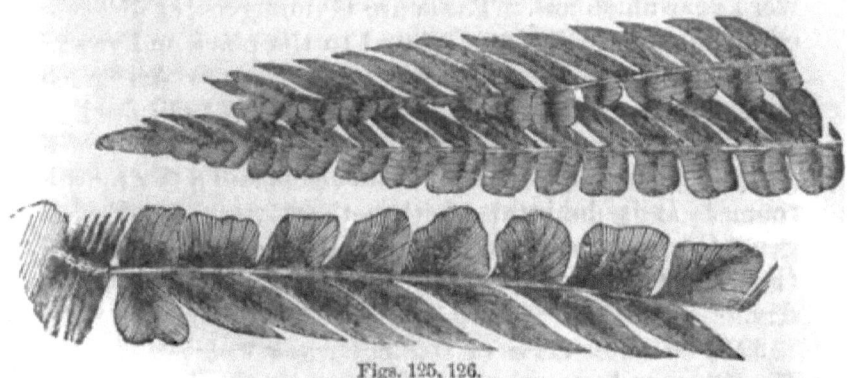

Figs. 125, 126.

127, and 128 (p. 224) are represented some fossil ferns. The two first are from the coal-measures of Rhode Isl-

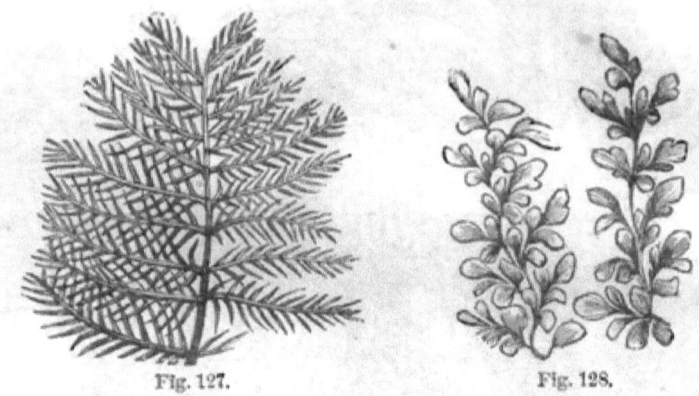

Fig. 127.   Fig. 128.

and. Many of the ferns were trees resembling the tree-ferns now found in some tropical countries, growing to the height sometimes of even 40 and 50 feet. It is stated by Hitchcock that over 250 species of ferns have been found in the coal strata of Europe, and yet the native species in Europe at the present time do not exceed 50. As ferns are now far more abundant in tropical than in temperate regions, the life-record of the coal-measures seems to show that the climate in Europe and this country during the coal period was tropical.

324. **Conifers.**—The Conifers, or cone-bearing plants, were very abundant. These are Gymnogens, or dicotyledonous plants (Fig. 87), related to the pines and yews. Most of the other plants of this age were Acrogens (Fig. 86), the Ferns, the Equisetaceæ, or cat-tail family, and the Lycopodiaceæ, or club-mosses. The Amphigens (Fig. 85) were represented in sea-weeds and a few mushrooms. It is doubtful whether there were any Endogens (Fig. 88), and there were none of the Exogens (Fig. 91), the chief forest-trees and shrubs of the present day.

325. **General View of Carboniferous Vegetation.**—In Fig. 129 you have grouped together many of the different kinds of plants of the Carboniferous age. On the right, in the foreground, is the Sigillaria, with Stigmaria

AGE OF COAL. 225

Fig. 129.

roots, and next to it the beautiful tree-fern. On the left, standing up loftily, is a Lepidodendron, and by it lies prostrate a Calamites, there being also in the rear of it, in the background, one inclining of a much larger size. The smaller plants fill up the intervals between these,

There were few, if any flowers in all this vegetation, but "what was wanting in blossom," says Page,* was more than compensated for by the profusion of light, symmetrical, feathery fronds, and by the tall, pillar-like stems which rose, each one boldly carved with its own peculiar pattern. The trunks of a modern forest are rough and gnarled; those of the period now under review sprang up like the sculptured shafts of a mediæval temple, graceful in proportion, and rich in ornament through the endless repetition of flutings, spirals, zigzags, lozenges, ovals, and other geometrical designs—these designs being the persistent leaf-scars of a vegetation simpler in structure and more primitive in plan." An example of the leaf-scars of which he speaks is represented in Fig. 130, the scars left by the fallen leaves being seen at $a$, while at $b$ is the surface of the wood brought to view by the removal of the carbonized bark. You observe the regularity with which these scars are arranged. The modes of this regularity differ in different plants, thus affording a rich variety.

Fig. 130.

326. **Causes of the Rank Carboniferous Vegetation.**—In order to understand how the rank vegetation of that age was produced, you must recur to what I said of the chemistry of the atmosphere in Part II. You there learned that from the carbonic acid gas which is supplied to the air from the lungs of animals, from fires, and from various chemical decompositions, the leaves gather in the most of the carbon which is incorporated in the plants in their growth. Now it is supposed that there was much more carbonic acid in the atmosphere up to the beginning of the coal-making age than there was at its close. Some estimate it as having been six times as

much as it is at the present time. Here, then, there was in the air a large surplus of one of the three materials of which plants are chiefly composed, and it was appropriated in producing the profuse vegetation. And as this vegetation was for the purpose of making coal, we may say that the material for the coal was kept in the atmosphere in a gaseous state, by being combined with oxygen, until the earth, during a lapse of long ages, was put into a fit condition to bring forth an exuberant vegetation. When this was effected the Creator introduced the requisite plants, and the transfer was made from the atmosphere through them to the coal-measures. Such a transfer from the gaseous to the solid state of such a quantity of matter seems at first thought strange, but it is no more so than the transformation which we see continually in the case of water. This rises in the vaporous or gaseous state into the atmosphere by evaporation, and anon we see some of it again in the shape of solid ice on the surface of the earth.

327. **Climate of the Carboniferous Age.**—The climate of the earth was very different in that age from what it is now. It was warm in all quarters, and very equally so. We know this because we find very widely distributed the remains of plants and animals similar to those which now flourish in warm climates. The climate was so warm in the arctic regions that rank vegetation was present there, and coal was laid down. With such prevalent warmth there must have been great moisture in the air, and this favored rank growth, notwithstanding there probably was from this moisture less of clear sunshine than there is now.

328. **Carboniferous Scenery.**—Though the vegetation of that age was so rich, the scenery was tame and monotonous. Forests, with their tangled undergrowth, mostly swampy, spread over vast platforms, in some cases almost continental in extent. There were no such elevations as we see at the present day to relieve the

eye, but one wide wilderness stretched far away, varied only by slight undulations. We may get some idea of the nature of this scenery from some localities on our earth, but none at all of its extent. Especially extensive were the platforms in the earlier part of the Carboniferous age, for the beds of **coal then laid down** are found to be much vaster in extent than those of the **later** coal-measures. In this country these early beds cover a large portion of the continent. Each of the successive platforms of luxuriant vegetation, after having maintained the same level for hundreds, or even thousands of years, was submerged in water gradually (Hugh Miller thinks in some cases suddenly), that each bed of coal might be packed down with layers of rock above it. The wide wilderness of vegetation thus became a wide waste of waters.

329. **Animals.**—So highly charged with carbonic acid gas was the atmosphere of the coal-making age, that animals which require such air as we now have on the earth could not exist. There were, therefore, no mammals or birds, and none of the higher kinds of insects. There were cold-blooded animals, as fishes and reptiles, the latter finding their appropriate localities in the stagnant pools, swamps, and thick, damp forests of that age. Cold-blooded animals live an inactive, lazy life, as compared with the warm-blooded, and therefore do not require so much oxygen.* There was, of course, an abundance of Mollusks, Crustaceans, and Radiates, especially when the platforms were submerged, and the limestone strata were forming. The tameness and monotony of

---

* It would seem at first thought that the life of fishes is generally a very active one; but it is not so, for the fish needs to make but little effort in its motions, for two reasons. First, it is of nearly the same specific gravity with the element in which it moves, and is therefore obliged to make but a slight effort in any upward movement. It is in strong contrast with birds in this respect. Secondly, most fishes have an air-bladder, which they can compress or enlarge at pleasure, as they wish to fall or rise. See my Natural History, Chapter xx.

the carboniferous scenery were heightened by the absence of all running and flying animals. As to sounds, "there was no music in the groves," says Dana, "save, perhaps, that of insect life and the croaking Batrachian."

330. **Coral Animals.**—The reef-making animals, some varieties of which you see in Fig. 131, were busy wher-

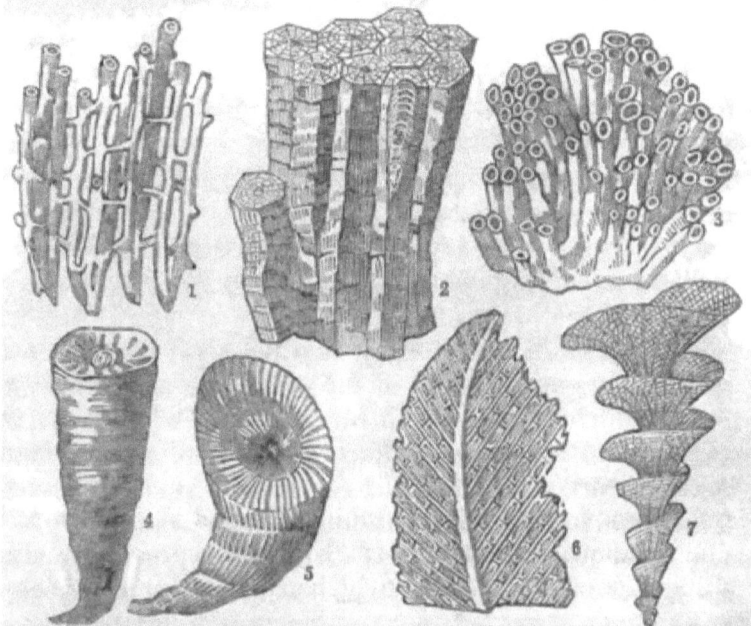

1. Syringopera. 2. Lythostrotion. 3. Aulopora. 4. Amplexus. 5. Clisiophyllum. 6. Ptilopora. 7. Archimedopora.
Fig. 131.

ever limestone was forming. They had a large work to do, then, in the sub-carboniferous period, when the great limestone floor was laid down for the coal-measures, and afterward, also, in the case of those strata over the coal-beds which are composed of limestone.

331. **Crinoids and Sea-urchins.**—These animals abounded in connection with the corals, the crinoids being much more abundant than the corals in the limestone of the sub-carboniferous period. Indeed, whole strata are composed of the calcareous remains of the crinoids. Two

varieties are represented in Fig. 132, at 1 and 2. There

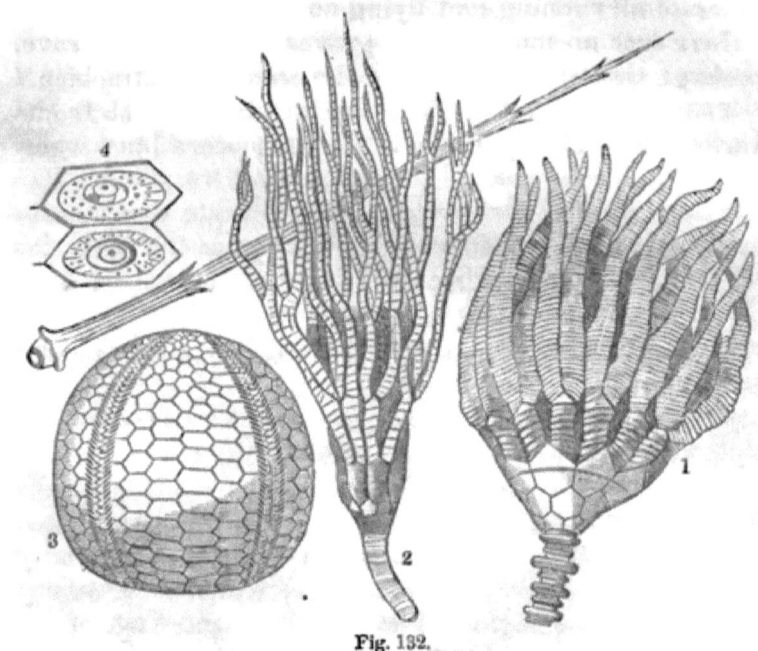

Fig. 132.

is also one variety of sea-urchin at 3, and the plates and spine of another variety at 4. Fig. 133 represents a slab of limestone from Iowa, in which are seen remains of cor-

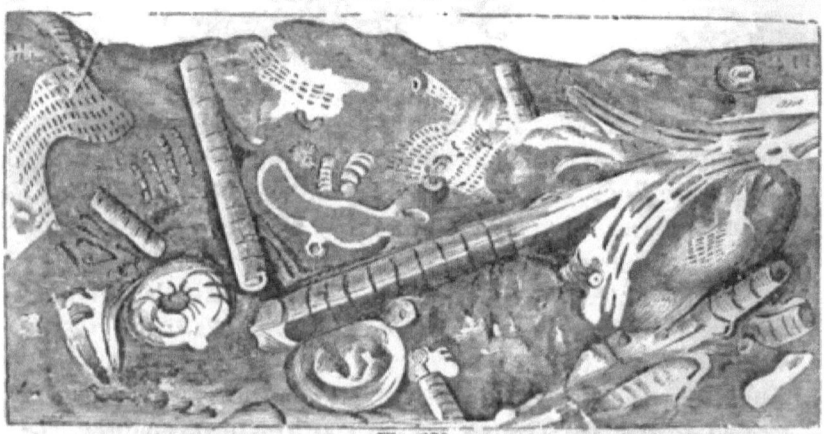

Fig. 133.

als, crinoids, etc., brought out in relief by the long-continued action of the weather upon the stone.

332. **Articulates.**—Of the Crustaceans, but few of those singular animals, **the Trilobites, which numbered about six** hundred species in the Silurian, were present in the Carboniferous age, and these were the last of that tribe. New tribes appeared, of which one was very much like the horse-shoe of the **present day.** Minute Crustaceans swarmed in myriads in stagnant waters, as their remains in the rocks now show. Of insects there have been found some wing-cases of beetles. Insects of the cockroach tribe flourished then, and there were also some scorpions.

333. **Reptiles and Fishes.**—Some reptiles appeared in this age, foreshadowing the succeeding one, in which they were to be so abundant and large as to give the name of Reptilian to it. The remains of some have been found, and tracks have been discovered which evidently belonged to reptiles, no remains of which have as yet appeared in the rocks. Fig. 134 represents two of the tracks of such a reptile, which were found near Pottsville, in Pennsylvania, by Dr. Isaac Lea, of Philadelphia. In one part of the slab there is the impression of a tail. These marks, mingled with ripple-marks and impressions of rain-drops, making out a most interesting record of that age so far distant in the past eternity, have been before noticed in § 210. Some of the fishes of this age foreshadow the succeeding Reptilian age, by combining some of the reptilian characteristics with those of fishes. Of such reptile fishes there is at the present time but one known genus, the Lepidosteus, or gar-

Fig. 134.

pike of North America. Of this Hugh Miller says that "it would almost seem as if it had been spared, amid the wreck of genera and species, to serve us as a key by which to unlock the marvels of the ichthyology of those remote periods of geologic history appropriated to the dynasty of the fish." This wonderful animal has an armor of bony scales, covered with hard enamel like that of teeth, so that it would be difficult for any shot to take effect on it; and as his teeth are very formidable, it has been said of him by the fisherman that "he can hurt every thing, and nothing can hurt him." It is very agile, darting through the water even up the rapids of Niagara. It can bend its head freely in all directions, like a serpent, which no other fish can do, for its vertebræ have the ball and socket arrangement of serpents instead of the cup arrangement of fishes. In this fish we have the only living type of the prevailing family of fishes in the Coal period. Some of the fishes of that family were such monsters in size, compared with the gar-pike, that the sublime language of Job about the Leviathan could be properly applied to them. Hugh Miller says of them, "If the gar-pike, a fish from three to four feet in length, can make itself so formidable from its great strength and activity, and the excellence of its armor, that even the cattle and horses that come to drink at the water's side are scarce safe from its attacks, what must have been the character of a fish of the same reptilian order from thirty to forty feet in length, furnished with teeth thrice larger than the largest alligator, and ten times larger than the bulkiest Lepidosteus, and that was covered from snout to tail with an impenetrable mail of enameled bone?"

334. **Mollusks.**—Of these there was a great abundance and variety. The Brachiopods were largely represented. Of the genus Productus there were many species, one of which is represented in Fig. 135. It is called Productus spinulosus, from the long, slender spines that project from the shell. Of another genus of the Brachiopods, Spirifer,

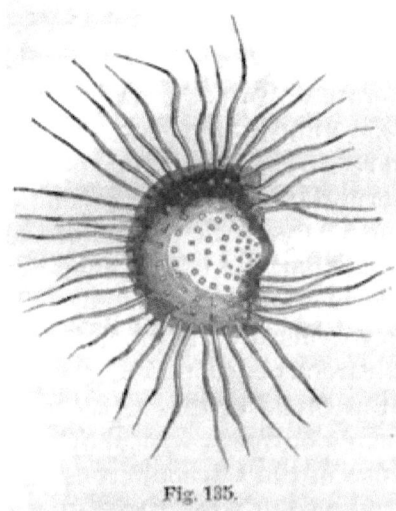

Fig. 135.

there were also many species. One of them is represented in Fig. 136, showing the spiral support from which it takes its name, *a*, as it is in the shell, and *b*, a part of it taken out. The strange Orthoceratites, which, like the Crustacean Trilobites, were abundant in the Silurian age, now, like them, appeared with but few species, and, before the age had passed, became extinct.

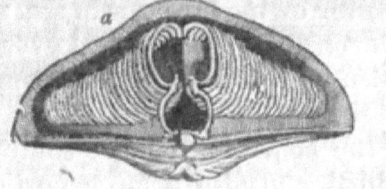

Fig. 136.

335. **Permian Period.**—The Carboniferous age ended with what is called the Permian period, in which no coal was formed. This period has its name from a portion of Russia which was in ancient times the kingdom of Permia. Here the rocks of that period cover an area 700 miles long and 400 broad. Until quite recently, it was supposed that there are no Permian strata in this country; but they have been discovered in Illinois, Kansas, and some parts of the slope of the Rocky Mountains. In this period nearly all the country east of the Mississippi was dry land, and therefore all the Permian deposits were made on the west of it, with the exception of a small part of Illinois. In Europe there are Permian strata in Central Germany, and in England in the neighborhood of the coal regions, in addition to the extensive area cov-

ered by them in Russia. In this country the rocks are limestones, sandstones of various colors, shales or marls, gypsum beds, and conglomerates. In the Permian strata of Europe the limestones are mostly found to be *magnesian*. Why this is so the researches of geologists have as yet not been able to determine; and Phillips remarks in regard to it that "we must be content to shelter our ignorance under the statement that, from some unknown cause, the waters of the sea were then decomposed in such a way as to permit very generally the precipitation of united magnesian and calcareous carbonates."

336. **North America at the Close of this Age.**—Two thirds of the North American Continent was above the level of the ocean at the close of the Carboniferous age. This was the eastern part, while the western was occupied by a large interior or mediterranean sea, such as in the Silurian and Devonian ages extended over much of the eastern part also. The Rocky Mountains, as well as, indeed, the Alleghanies, did not yet exist. The borders or fringes of the continent on the south and east were yet to be added, and also a large portion of the western part of the continent. Florida, that great work of the coral animals, was yet to be constructed. Although so much was done, long ages upon ages were yet to be passed before the continent would be completed.

337. **Disturbances in the Coal-measures.**—The strata of the Carboniferous age have been subjected to upheavals, foldings, fractures, etc., like all other strata, some of them being very extensive, even mountainous. If it were not for these, few of the beds of coal lying between the thousands (§ 318) of feet of strata would have come within the reach of man in his minings. Most of these upheavals and flexures occurred at the close of the age, in the interval between Palæozoic and Mesozoic time. The results are so observable in this country, all along the region of the Appalachian chain, that the change produced at that time has been called the Appalachian revolution.

Where the flexures are most decided the strata were metamorphosed—gneiss, granite, and other crystalline rocks being produced—showing that great heat was present at that time in such localities. The same changes took place in the coal-measures of other countries. The movements in the bending and folding of the strata were not confused, but occurred in certain lines, according to some systematic plan of the Creator. They were probably very slow—perhaps to the extent of some yards, or even feet only, in a century—else such bendings without fractures as are often found could not have been produced. It is believed that, besides the metamorphosis of strata which laid up at that time for the use of man such quantities of building material, there were separated and deposited in this metamorphic process many of the precious and other metals, and also many of the precious gems. It has also been observed that the coal is most debituminized where the disturbances of the strata were the greatest. This is probably to be attributed to two causes— the action of great heat in such localities, and the opportunity of escape for gaseous products given by the fractures made in the disturbances. Denudation is often connected with the faults, bendings, etc. This is exemplified in Fig. 68, p. 140. These denudations were sometimes vast, the thickness of strata removed amounting even to many thousands of feet.

338. **Palæozoic Life-record.** — You have now been made acquainted to some extent with the changes in life which occurred in the three Palæozoic ages—the Silurian, Devonian, and Carboniferous. While the grand divisions marked out by the Creator are preserved from the beginning, showing that the same general plan existed from the outset as now, the predominant forms of life, especially of animal life, are very different from each other in the different ages, and very different from those which appear at the present time. All the four divisions of animals began in the first of the Palæozoic ages, the

Silurian; but the vertebrates did not appear till the latter part of it, and then in small numbers, and only in the low form of fishes. Reptiles did not appear till the latter part of the Devonian, the age of Fishes, and were not in their greatest abundance till after the Palæozoic time was past. There were no Mammals in the Palæozoic ages. Many species of animals which flourished in those ages, perhaps largely, at length vanished, never to appear again; for it is a remarkable fact, that any species which comes to an end is never again brought upon the stage. For example, the Trilobites, of which there were hundreds of species in the Silurian age, all dwindled away one after another, till, in the Carboniferous age, the last of the species disappeared, and none of all of them has ever again been revived. At the close of the Carboniferous age, in the midst of the disturbances that occurred, every species of animal and plant was destroyed, though many species belonging to the same genera with those of this age existed in the after ages. This universal destruction of the life of the Carboniferous age is one of the great marks of distinction between this and the following age.

## CHAPTER XVII.

### AGE OF REPTILES.

**339. Names.**—We now pass from Palæozoic to Mesozoic time, or the Mediæval age of the earth, the age of Reptiles. It is divided into three periods, the Triassic, Jurassic, and Cretaceous. The Triassic is so called on account of an obvious threefold division which the system of rocks presents in Germany, though it is not seen in other quarters of the world. The term Jurassic comes from the Jura Mountains of Switzerland, where the formation or system of strata designated by this name is well developed, and has been particularly examined.

The term Oolitic is sometimes applied to it, though commonly it is confined to one portion only of the Jurassic period, in which one of the limestones formed is oolitic in its structure. The term Cretaceous, from the Latin word *creta*, chalk, is applied to the closing period of the Reptilian age, because the chalk of England and Europe is one of the rocks of the system belonging to that period. In this country we have the system, but the chalk, which is so prominent in those countries, is left out here.

340. **Triassic Rocks.**—The rocks of this period are mostly sandstones, shales, conglomerates, and sometimes some limestones. So prominent are the red sandstones that the whole system is called the New Red Sandstone, in distinction from the Old Red Sandstone of the Devonian age. In the valley of the Connecticut the material for the rocks of this formation was derived from the crystalline rocks in the neighborhood, chiefly by the erosive action of water. The freestone of Connecticut and New Jersey, so much used in building, comes from this formation. In the western part of this country the Triassic rocks are sandstones and marls of a brick-red color, often having gypsum in them. The red color, so prevalent in the sandstones, though varying much in shade, is owing to the presence of oxyd of iron. The life-record in the rocks of this formation is scanty, because they are not well suited to the preservation of fossils.

341. **Localities.**—In the eastern part of this country the Triassic rocks are surface rocks along between the Appalachians and the Atlantic coast. In the Connecticut Valley they extend from the Sound at New Haven up to the northern part of Massachusetts. There stand up in the midst of them Mount Holyoke, Mount Tom, East and West Rock, and other Trappean elevations, imparting variety and grandeur to the scenery. In the West this system is extensively spread over part of the slopes of the Rocky Mountains. The Triassic system is developed in various parts of Europe and in England.

342. **Triassic Salt.**—While in this country the chief source of salt is in the Silurian formation (§ 278), in England and Europe it is in the Triassic system, which is often, for that reason, called the Saliferous system. The rocks in connection with which the salt is found are similar in both formations, but the manner in which the salt was produced could not be the same in both cases. The evaporating process which it is supposed, as stated in § 278, produced it in the Silurian formation in New York, could not in any way accumulate masses of rock salt 40 yards in thickness, or a mountain of salt 600 feet high and 1200 broad, such as is found at Cordova, in Spain. It is supposed that volcanic agency operated in such cases, as salt is common in what is thrown out from volcanoes, and salt-springs sometimes rise to the surface from granitic rocks, showing that there are sources of salt lying deep in the earth. The purity of the rock-salt, it is thought, indicates its volcanic origin, for if it were deposited from a solution it would have impurities mingled with it.

343. **Plants of the Triassic Period.**—Though the plants which were peculiar to the Carboniferous age were not present, as the Sigillaria and the Lepidodendrons, yet there were Cycads as there were then, and Ferns, Equiseta, and Conifers in new forms. Trunks of conifers of considerable size have been met with in the sandstones. Some coal has been found in this formation in some localities, made, of course, out of the vegetation of which I have spoken. Triassic coal has been discovered in Virginia and North Carolina in this country, in Australia, and in various parts of Asia.

344. **Animals.**—The excess of carbonic acid that existed in the atmosphere previous to the Carboniferous age was removed from it by the rank vegetable growth of that period, so that when the Triassic period came on the air was fitted for air-breathing animals. They accordingly were introduced upon the scene. Not only

were reptiles, and among them the monstrous Saurians, introduced, but also birds and mammals. The mammals were not, however, in great abundance, and were of the lower orders, the marsupial, which at the present day are so common in Australia. The fishes were all ganoids and placoids, as in the Palæozoic ages, but there was an approach in their tails to the bilobed character of the present day (§ 302), and some were wholly of this modern fashion. Of the Mollusks, a family called Ammonites, which figured largely in the two other periods of the Reptilian age, now appeared. One species, the Ammonites 'nodosus, is represented in Fig. 137. Among the radiated animals of this period was a crinoid of singular beauty, called the Lily Encrinite, Fig. 138.

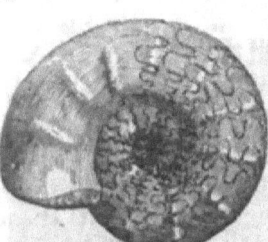

Fig. 137.

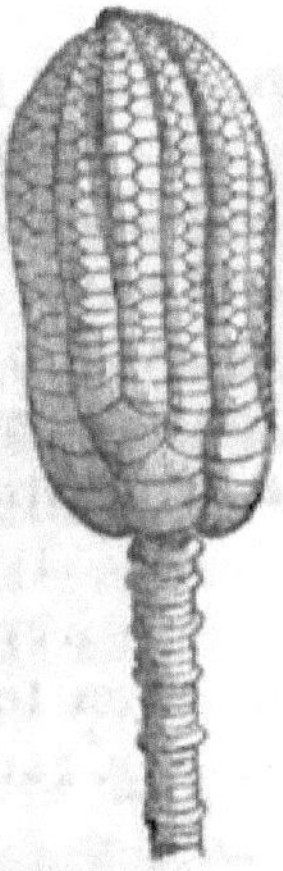

Fig. 138.

345. **Reptiles.**—Lizards, Saurians, Batrachians, many of them gigantic in size, were abundant, though not so abundant, and strange, and monstrous as they were in the following period, the middle one of the age of Reptiles. One of the most remarkable of these animals was the Labyrinthodon, pictured in Fig. 139 (p. 240). Though having a head of three or four feet in length, and teeth three inches long, and being about the size of an ox, with his long hind legs, he was very much like a frog. All of his skeleton has never been discovered, but from his skull and some other bones Professor Owen, by his knowledge of the adaptation of bones to each other, like Cuvier, has reconstructed in his mind the whole frame of the animal, and the result is the singular form which you have in the figure. The tracks which you see in the figure have been observed alone, in Triassic rocks, and have, until recently, been supposed

240  GEOLOGY.

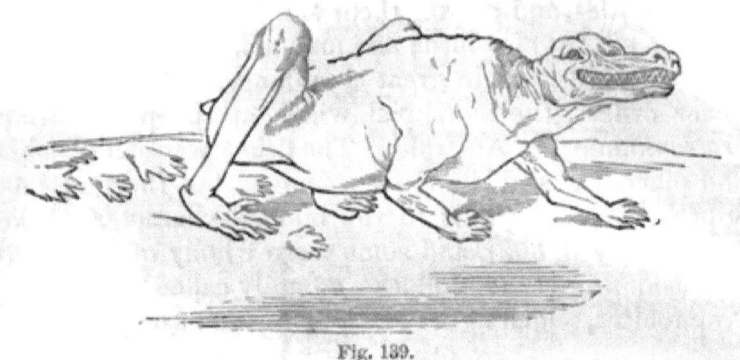

Fig. 139.

to belong to another animal, whose remains, however, had never been discovered. The name of Cheirotherium was given to this supposed animal because the tracks were so much like the print of a hand, the word being derived from two Greek words, *cheir*, hand, and *therion*, beast. It is now pretty well ascertained that these tracks were all made by the Labyrinthodon. This name is given to this animal from the structure of its teeth. In Fig. 140, at *a*, is a tooth half its natural size, and at *b*

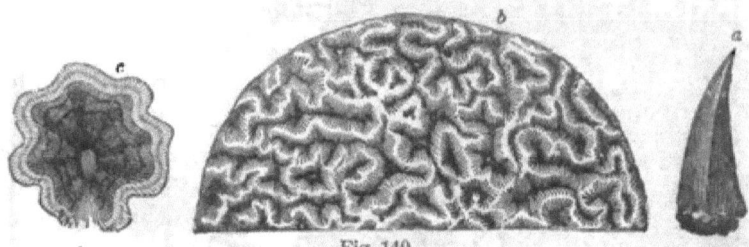

Fig. 140.

is part of a transverse section magnified twenty diameters. You see that the turnings on this surface have a labyrinthine arrangement, and hence the name given to the animal. At *c* is one of these turnings very highly magnified.

346. **Tracks.** — The tracks of Triassic animals have been observed by geologists in various quarters, but most largely by Professor Hitchcock, of this country, in the red sandstones of the Connecticut River Valley, where

# AGE OF REPTILES.

they abound. Before he began his observations in 1835, these tracks had been noticed for full forty years by persons who were not aware of their great geological interest. Some are tracks of birds, some of reptiles, and some of animals that had in combination the characteristics of both birds and reptiles—reptilian birds, as they may be called. In Fig. 141 is represented a slab of sandstone found near Mount Holyoke. The large tracks show that the animal which made them on the mud, now hardened into rock, had a foot 20 inches long and 21 wide. It is supposed that he was a Batrachian, a toad-like animal, and yet he must have been of the size of an elephant. There are tracks, *a b*, of some smaller animal, and much of the surface shows also the impressions of rain-drops as distinctly as if they were made yesterday. One of the largest of the birds of that period was the Brontozoum giganteum, whose track is given in Fig. 142 (p. 242). Its feet were from 14 to 20 inches long, its

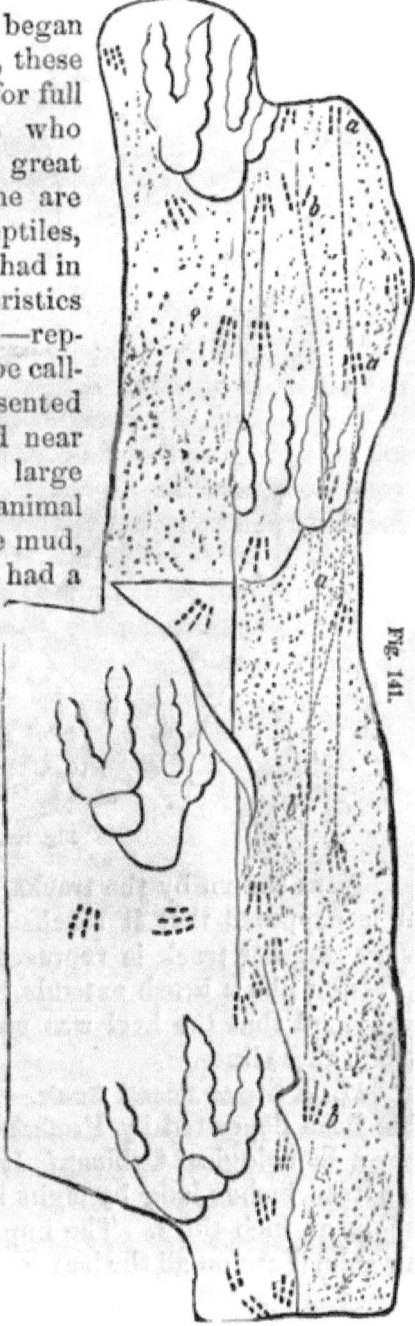

Fig. 141.

L

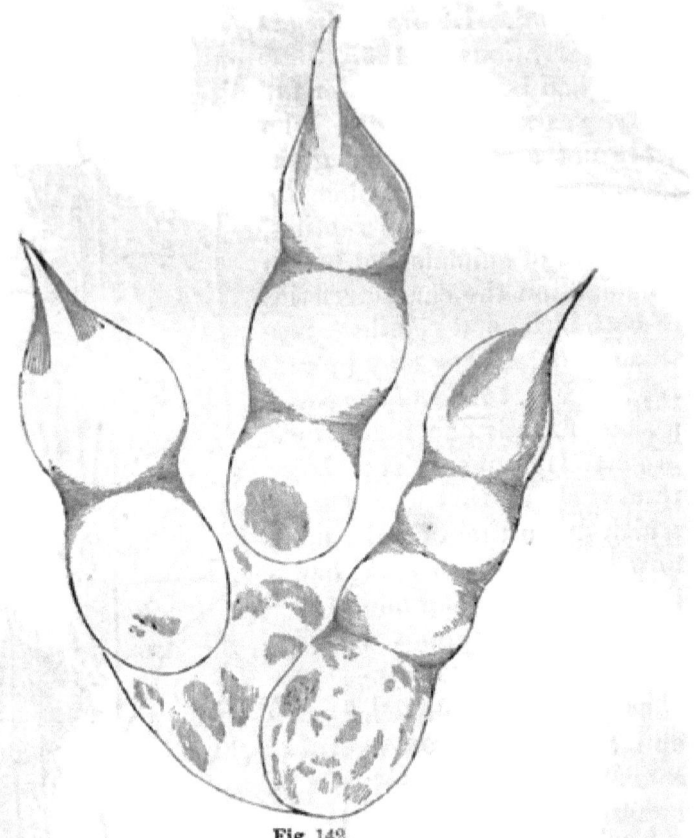

Fig. 142.

stride, as shown by the tracks, from four to six feet, and it is supposed that it reached a height of 14 feet. A very singular track is represented in Fig. 143. An impression like a brush extends back from the foot. It is supposed that the heel was covered with ridges, which made these lines.

347. **A Stone Fossil Book.**—A book with stone leaves has been deposited by Professor Hitchcock in the Amherst Ichnological Cabinet. It is a book of five leaves, nineteen inches long by eight in width, each leaf being about an inch thick. The impressions of the tracks of an animal are on all the leaves, corresponding in position

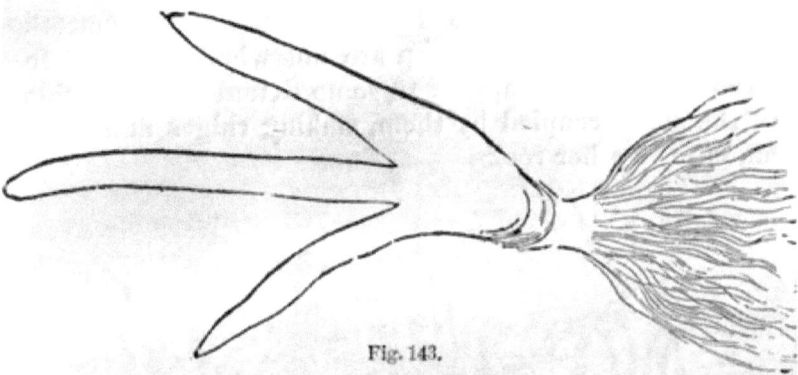

Fig. 143.

throughout. The explanation is this: Five successive layers of mud had been laid down of this thickness, each one having partly dried before the succeeding one was laid. The animal treading upon the upper one made an impression through the whole five layers, which afterward solidified into rock, and so were preserved through long ages to the present time. Dr. Hitchcock ingeniously mounted the leaves in such a way that the book can be readily opened or shut.

348. **Trap Rocks.**—Masses of these rocks have come up among the Triassic rocks in the eastern part of this country, many of them being lofty enough to be called mountains. As examples of these, there are, in Massachusetts, Mounts Tom and Holyoke; in Connecticut, East and West Rock at New Haven, and the Hanging Hills of Meriden; in New Jersey, Bergen Hill, and various other elevations. I have already told you, in § 230, how these masses are formed. The effect which the fused rock, as it was thrust upward, produced upon the sandstone is plainly to be seen in many localities. The difference between the sandstone close by the trap and that which is away from it is as distinct as that between a brick so burned as to be very hard and one which has the ordinary hardness. The ridges and dikes of trap are arranged with some system, extending in certain general directions as mountains do. This may be seen in Perci-

244　GEOLOGY.

val's map of them in his Geological Survey of Connecticut. The eruptions of trap are not wholly confined to Triassic rocks, but appear to some extent on either side of the area occupied by them, making ridges and dikes among the other rocks.

Fig. 144.

**349. Plants of the Jurassic Period.**—The climate of the

earth was quite universally warm and genial, as is shown by the wide diffusion of certain plants and animals, even into the arctic regions. The vegetation, therefore, was luxuriant, consisting largely of cycads, conifers, and tree-ferns, with some palms, lilies, etc. There were also equisetums and club-mosses. In Fig. 144 is represented the vegetation of this period. In many localities the luxuriant vegetation was packed down in beds of coal, as in the Carboniferous period. There are localities of this kind in Europe, in Great Britain, in China, in India, and in Virginia in this country. Coal can be made of vegetable fibre at any period if those circumstances are present which I have stated to be necessary in the chapter on the Coal-making age.

350. **Portland Dirt-bed.**—In the island of Portland, and in some of the adjoining counties of England, there have been found the remains of an ancient forest of the Oolitic, or Jurassic period, lying between the strata. This so-called Portland dirt-bed is composed of a bluish loam, which contains stumps and roots of trees petrified, as represented in Fig. 145. There are strata of rock

Fig. 145.

above and below; those below, A, being mostly marine limestone, and those above, C, being, on the other hand, fresh-water limestone. The rocky strata and the dirt-bed, you see, are all parallel, but not horizontal. They were horizontal when they were formed, but have all in some way been tilted up together. This tilting, however, is only found in some localities, the horizontal position being retained in others. Now, in all the region where this arrangement is found, there must have been, in the first place, deposits of a marine character, for the Portland stone, A, is filled with marine shells. Afterward these beds were covered with lake or river mud, which became dry land, and was the scene of a forest vegetation. Then the land sank down, and was submerged, with its load of forest growth, beneath fresh water in place of the salt water which was there before the dirt-bed was laid. From this water the limestone strata, C, were deposited upon the dirt-bed, B, this being demonstrated by the shells in the stone, which are of the kinds that are found only in fresh water. The tilting, which is to the amount of 45 degrees, was not done till all the fresh-water limestones, C, were laid down.

351. **Animals.**—Animal life had advanced on its forms in the previous age, though the fishes were still only ganoid and placoid, and the mammalia were still of the marsupial type. Insects appeared more largely than before, because there was more food for such animals in the vegetation. "There were," says Page, "burrowers among the decaying timber of the pine forests; leapers among the leaves and herbage of the cycas grove; hunters along the river-bank and across its sunny waters; and gaudy flutterers over the flowers of the lily and the palm-tree. All the great orders of insect life—beetles, cockroaches, dragon-flies, grasshoppers, and ants—are abundantly represented." The Crustaceans were numerous, those which were similar to shrimps and lobsters being especially prevalent. The higher orders of Mollusks abounded. There were still Crinoids, though they were on the wane. Corals were abundant wherever the formation of limestone was going on. Sea-urchins also were numerous, one of which is represented in Fig. 146. But the reptilian development was the great feature of this middle portion of the age of Reptiles. There were reptiles for the ocean, the river and estuary, the muddy shore, and the land;

AGE OF REPTILES. 247

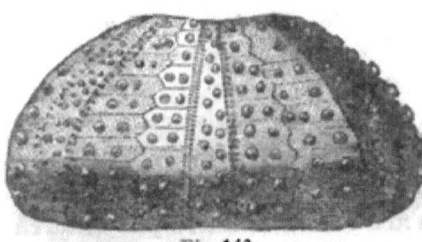

Fig. 146.

and many of them were monstrous in size and extraordinary in character.

352. **Ammonites and Belemnites.** — Of the Mollusks, these are the most prominent and peculiar. They belong to the class called Cephalopods (§ 286). The Ammonites, which began in the Triassic period, were now very abundant, their beautiful shells

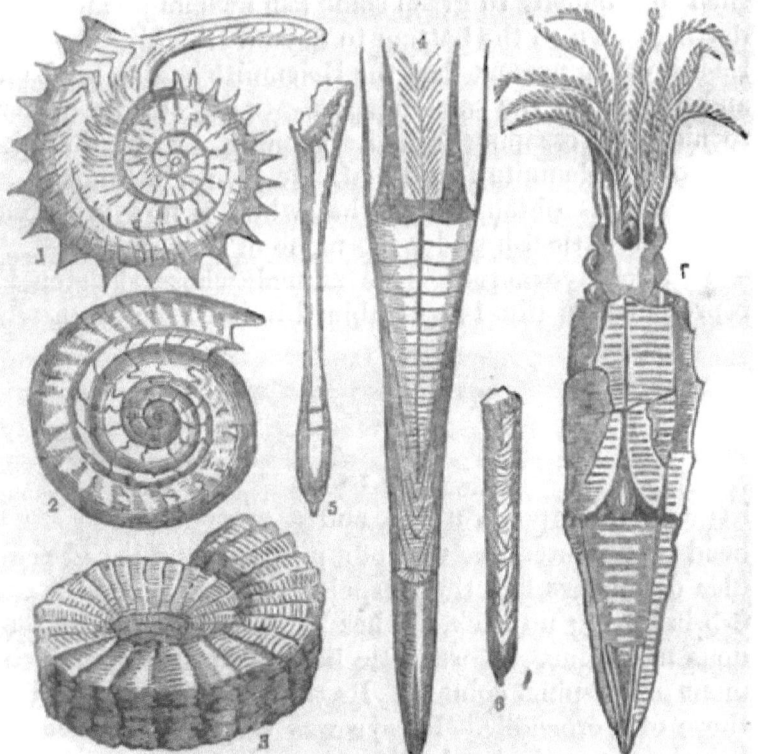

1. Ammonites Jason. 2. A. Communis. 3. A. Bucklandi. 4, 5, 6. Belemnites. 7. Belemnoteuthis.

Fig. 147.

appearing in hundreds of forms, with every variety of

ornamentation. It was now the "reign of the Ammonites;" but they have long ago passed away, and the only present representative of the class is the Nautilus of the Southern Ocean. Some of these ancient Ammonites were monsters, being even three feet in diameter. On the left of Fig. 147 (p. 247) are three species. On the right of the figure are the Belemnites, in different degrees of preservation. These belong to the Cuttle-fish tribe. At 7 is one with its crown of arms, with which it took its prey. The animal probably sailed along with its long, conical shell pointed downward, and now and then rose quickly to grasp some fish swimming above it, darting down to the bottom to devour it. Like the cuttle-fish of the present day, the Belemnite had an ink-bag, and used it for the same purpose—to discolor the water to aid it in escaping from its enemies. From the ink-bag of a Belemnite found at Lyme Regis, in England, a pigment was obtained like that which is now prepared from the cuttle-fish under the name of India ink.

353. **Ichthyosaurus.**—This animal, whose skeleton is represented in Fig. 148, combined in itself the character-

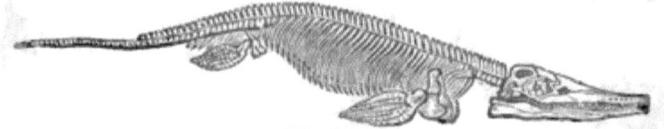

Fig. 148.

istics of a porpoise, a fish, and a crocodile. Its large head was pointed like that of a porpoise, and it had paddles or flippers like the Porpoise family. It was like a fish in having no neck, the head and body being continuous in outline. It was also like a fish in the arrangement of its spinal column. Its teeth and jaws were like those of a crocodile. Its eye was very large, its socket being eighteen inches in diameter. It was so constructed that the animal could see equally well in and out of the water. Its jaws were so long that, when fully opened, their extremities would be seven feet apart. It was, in some cases, even thirty feet in length. One of its paddles contained over a hundred bones, giving this instrument great elasticity and power. Thirty species of this animal have been discovered.

354. **Plesiosaurus.**—This animal, of which you see the skeleton represented in Fig. 149, had the general plan of

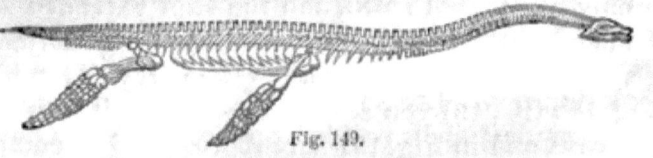

Fig. 149.

the Ichthyosaurus, but differed from it in several material particulars. It was fitted for more brisk motion. Its head was flat and serpent-like. It had a very long neck, its paddles were large but slender, and its tail short. In swimming, it probably used its paddles more and its tail less than the Ichthyosaurus. The greatest difference, however, is in the size of the head. The length of the Plesiosaurus was about 17 feet. Both animals had the power of creeping on land.

355. **Pterodactylus.**—The Pterodactyl is the most extraordinary of the animals of this period. It had the head and neck of a bird, the mouth of a crocodile, the

Fig. 150.

wings of a bat, and the body and tail of a mammal. It was, in some cases, of such enormous size that it measured from tip to tip of the spread wings 18 or 20 feet. Although the general appearance of its wings was like the bat, the arrangement was different. Instead of four extended toes enveloped in skin, as in the bat, the fifth toes only were lengthened, and the skin extended along the side of the body and legs. Its eyes were enormously large, so that it probably could seek its prey in the night. In Fig. 150 (p. 249) you see the skeleton as it was found in one case, and in Fig. 151 it is represented as complete

Fig. 151.

and in order, so that you see the relative position and size of the various parts.

356. **A Question in Comparative Anatomy.**—In Fig. 152 we have the Ichthyosaurus, Plesiosaurus, and Pterodactyl as they are supposed to have appeared in the scenes of the Jurassic age. It has been commonly sup-

Fig. 152.

posed that the Pterodactyl was a flying animal, as here represented, and that it lived on insects as the bats do, although, perhaps, at the same time, it would dive down occasionally into the water for fishes, to mingle them with its insect food. Agassiz expresses, on the contrary, the opinion that it was an aquatic animal, preying upon fish and other animals in the water. This opinion he bases upon two facts. First, the teeth are conical, sharp, and separated from each other, and are, indeed, such as are common to animals that live on aquatic prey, and such as would not be required for the capture and destruction of insects. Secondly, there is nothing in the frame-work of the breast of the animal to indicate that it had those powerful muscles which flying animals always have. Instead of the projecting keel which birds have for the attachment of their large flying muscles, it has a thin, flat breast-bone, like that of the sea-turtle of the present day. For the full exposition of the subject of arrangements for flying, I refer you to my Natural History, §§ 199 and 200. It may be added to the reasons which Agassiz has given for his opinion that there were, so far as we know, no insects of sufficient size to be food for so large an animal. But, after all, there was in the Pterodactyl an apparatus which might be used, to some extent, in flying, though there were no muscles competent to sustain flight for any length of time. We must therefore conclude that it probably rose into the air occasionally, very much as the flying-fish now does, with, perhaps, a somewhat longer range of flight.

357. **Dinosaurs.**—There is a class of Jurassic land reptiles called Dinosaurs, the name coming from two Greek works, *deinos*, terrible, and *sauros*, lizard. Two of these, Megalosaur and Hylæosaur, are represented in Fig. 153.

Fig. 153.

Another is the Iguanodon, so called because it was like the Iguana family of the present day, a notice of which you will find in my Natural History, § 326. These monstrous animals, being 25 or 30 feet in length, roamed, elephant-like, over the river plains or through the forests, the Iguanodon browsing upon shrubs and trees, and the Megalosaur and Hylæosaur devouring crocodiles and tortoises.

358. **Cretaceous System.**—In Europe and Asia the series of rocks belonging to this system is characterized by the presence of sands and sandstones in the lower part, and chalk in the upper part. In this country there is no chalk in the series, but there are beds of sand, clay, marl, and different kinds of limestone. Sandy strata of various kinds and colors predominate. Some are solid, and others are more or less loose in their structure—some so much so that they can be crumbled by the hand. There is a dark green variety called the *Green-sand*, the greatest part of which, sometimes even to 90 per cent., is silicate of iron and potash. There is also a little phosphate of lime in it. It is prized as a fertilizer in New Jersey and other parts of this country, its virtue in this respect coming from the potash and the phosphate. In England the Green-sand gives its name to the whole of the lower part of the Cretaceous series, because it is so prominent in it. Chalk, which is so prominent a part of the series in England and Europe, is, like limestone and marble, a carbonate of lime, but loosely put together in comparison with them. The chalk is varied in character by the different kinds of impurities that are apt to be mingled with it, and their different proportions.

359. **Localities.**—In this country the Cretaceous system comes to the surface over a vastly larger area than the other two series of Mesozoic rocks, the Triassic and Jurassic. It, in fact, made a very large addition to the North American continent in the west and southwest, chiefly west of the Mississippi, though quite largely in the states bordering on the Gulf of Mexico. In the east it appears here and there from New Jersey down to

South Carolina. It is very prominent in England, the famous Dover Cliffs being composed of the chalk, and it appears on the other side of the Channel, in France. It also shows itself in other portions of France, and in various parts of Europe.

360. **Source of the Chalk.**—Chalk is a marine deposit, and it is supposed that it was made in water some two or three hundred feet deep. It was made chiefly by those minute animals called Foraminifera, the shells of some of whose species are represented in Fig. 83, p. 157. The evidence on this point is decisive. The observations of Ehrenberg and others with the microscope show that chalk is made up of shells, the foraminifera furnishing by far the largest proportion of them. Ehrenberg states that a cubic inch of chalk often contains over a million of these microscopic organisms. Whenever you make a mark with chalk upon the blackboard, you really deposit there a quantity of the shells of these organisms; and if your eyes could be suddenly endowed with a high magnifying power, that white line would appear to you like part of the wall of a grotto covered over with shells. Indeed, the fact that the carbonate of lime has in any case the form of chalk is owing to the aggregation of these shells, they being so small that they give to the rock formed from them a soft and porous character. This character is not given to the rock under any ordinary circumstances when it is formed from corals or shells of any size. These same foraminifera are busy at the present day taking from the water of the ocean carbonate of lime to form their shells, and these, being left on the sea's bottom, have been found in some localities to compose almost wholly the sand that has been brought up in deep soundings. The material for chalk is then being deposited, and the chalk which may thus be made may at some future period be raised to the surface, and constitute a part of the dry land of our earth.

361. **Flint in the Chalk.**—In England, and other coun-

tries where chalk abounds, flint is found in it to a considerable depth. Sometimes it appears in layers, but commonly in nodules or lumps, of various sizes and shapes. The nodules are sometimes irregular and grotesque, and when so are quite a favorite ornament in cottage gardens in England. Covered with whitewash, they are used as edges to the borders. One of these forms you see in Fig. 154. Sometimes these nodules

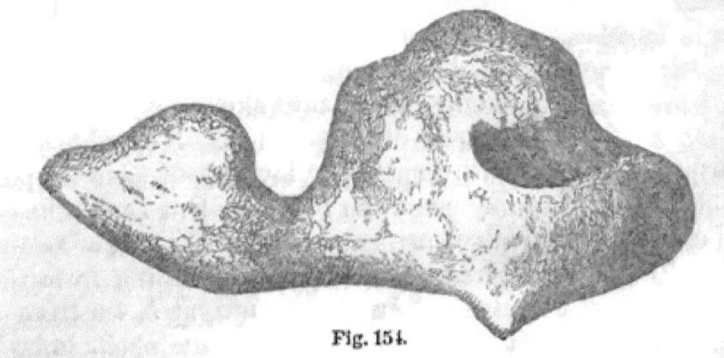

Fig. 154.

take such a form as to be called fossil mushrooms. In

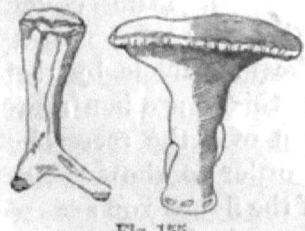

Fig. 155.

such cases the flint was deposited upon and in certain compound polypes of these forms, or, in other words, they are petrifactions of the frame-work of these animals. Two of these are seen in Fig. 155. Such forms are called Ventriculites.

362. **Fossils in the Flint.**—There are some very interesting minute fossils in the flint nodules. Some of these I will notice. There are certain vegetable fossils called Xanthidia, which, like the diatoms, were formerly supposed to be animal. There are many species of them. In Fig. 156 (p. 256), at 2, you see represented a chip of flint, which is so thin that the Xanthidia can be seen by transmitted light. The five spots mark the frame-works of five of them. The figure below exhibits one of these

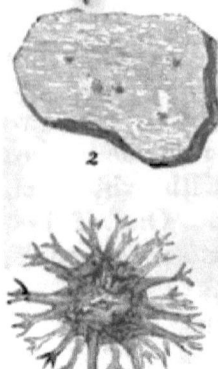

Fig. 156.

largely magnified. It is supposed that each of them furnished a nucleus around which the flint, dissolved in the water of the ocean, gathered and solidified. In the upper part of Fig. 157, at $a$, $c$, and $d$, are shell prisms that are found in flint, those at $a$ being the most common form. They are magnified ten times, as indicated.* The origin of these prisms, which are composed of carbonate of lime, is curious. They come from the shells of certain bivalves of that period. Now shells are ordinarily made with laminæ or layers, one placed upon another, as you may see in an oyster. But the shells of these strange bivalves are composed of prisms, or many-sided columns packed together, and extending from the inside to the outside. In the figure, at $b$, we have a piece of flint, with a piece of one of these shells imbedded in it. Some of the shell has been dissolved and removed, so that you see in the recess the little columns standing up all around, and on the floor of the cavity are the minute depressions in the flint made by their ends. In the specimen of which this figure is a representation there was a roof of flint over the recess, but this was carefully ground off in order to show the arrangement. In the lower part of the figure you see represented a great variety of flinty spicula which come from sponges, and are found in the flint nodules. They are all much magnified, one of them, $i$, even a hundred times. This is very rarely found. The most common of these spicules are $g$, $h$, $k$, and $p$. The spindle-shaped one, with raised rings, $o$, is by no means uncommon. The one resembling it, but having a triradiate summit,

---

* Throughout Fig. 157 the character × means multiplied or magnified, the numeral annexed showing to what degree the object is magnified.

# AGE OF REPTILES. 257

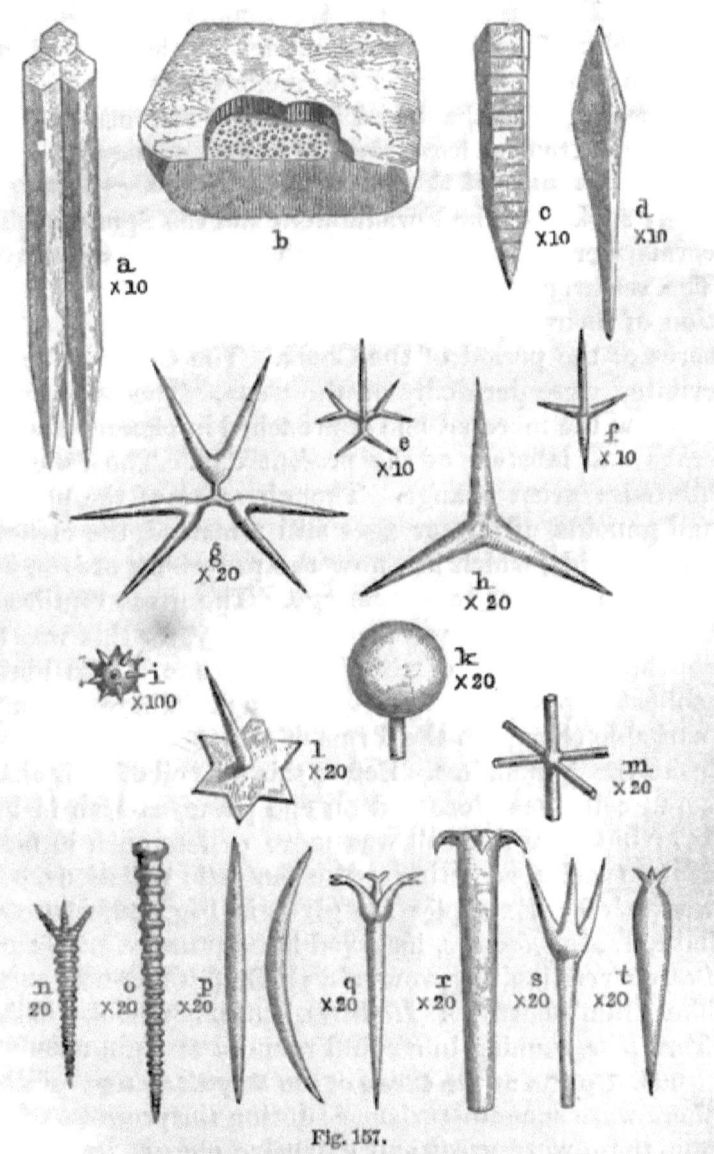

Fig. 157.

*n*, occurs more rarely. At *m* is the form which prevails in patches of silicious network. It is supposed that the spicula of sponges and the shells of the little diatoms furnished a large portion of the flint for the nodules in the chalk. The sponges abounded at that period, and had an agency in regard to silicious matter similar to that which corals have had in relation to limestone, and, to some extent at least, they have that agency still.

363. **Animals of the Cretaceous Period.**—I have already spoken of the Foraminifera and the Sponges. The corals were not so abundant as in the Jurassic period. The sea-urchins were very numerous, and the preservation of their beautiful remains is one of the marked features of the period of the Chalk. The Crinoids, or Encrinites, were decidedly on the wane. The Crustaceans were on the increase, and approached in character to the crabs and lobsters of the present day. The fishes exhibited a great change. Though some of the placoids and ganoids of former ages still remained, the ctenoids and cycloids, which are now the prevailing orders, first appeared in the Cretaceous age. The great reptiles of the Jurassic period were passing away, for this was the concluding period of the Reptilian age. The higher mollusks appeared in great profusion. There was a remarkable change in the forms of that great tribe of mollusks, the Ammonites. Before this the coil of their chambered cells was close and on one plane, as seen in Fig. 147; but now the coil was more or less open in many of the species, sometimes with fantastic variations, or it was spiral. Examples are given in Fig. 158, where we have, 1. *Ancyloceras*, incurved like a crosier. 2. *Scaphites*, curved like the prow of a skiff. 3. *Crioceras*, curled like a ram's-horn. 4. *Hamites*, bent like a hook; and, 5, *Turrilites*, running in a spiral round a straight axis.

364. **Uplifts at the Close of the Reptilian Age.**—While there were some disturbances during the progress of this age, there were great and extensive ones at its close, in the interval of passage from Mesozoic to Cainozoic time, just as there were at the close of Palæozoic time, as noticed in § 337. Some of the great chains of mountains

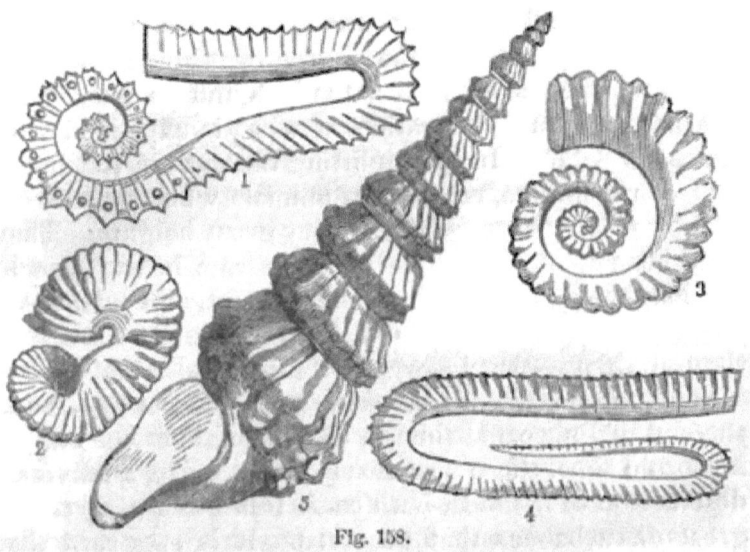

Fig. 158.

rose into existence in this interval—the Rocky Mountains, the Andes, portions of the Alps, etc. The same uplifting continued into the next age, so that these mountains, and the Himalayas, Pyrenees, Appenines, etc., were then fully raised. The evidence in regard to the rising of mountains during both these periods is twofold. 1st. The rocks of the cretaceous system lie high up on their sides; 2d. The marine rocks of the tertiary system—that is, of the age of Mammals, flank their sides below at a different slope. Take, for example, the Rocky Mountains. The cretaceous rocks are found lying sloped high up upon them. This proves that the sea, from whose waters the materials of the cretaceous rocks were deposited, stood over the region where these mountains now are, and that these rocks were raised up after the Cretaceous age had passed. Then the tertiary rocks were deposited from waters which flowed about the base of the mountains. But how do we know that the raising of them was not completed till some time in the Tertiary age? If it was completed, the rocks of the tertiary system would, of course, not be sloped up the sides of the mountains at all; but if it was not, they would be thus sloped in proportion to the degree of raising in the tertiary, and such sloping is actually found to exist. All this is simply an illustration of what is explained with figures in § 226. In this uplifting of strata in the formation of mountains, remains of animals which once lived in the sea have been raised to very great heights. They have been found imbedded in the strata in the Alps at the height of six or eight thousand feet, and in the Andes, according to Humboldt, at the height of fourteen thousand feet—that is, over two miles and a half.

365. **Destruction of Life at the End of this Age.**—As at the end of Palæozoic time (§ 338), so now at the end of Mesozoic time, there was nearly, if not quite, a universal destruction of life. In both cases this was owing to the great disturbances that occurred. It is supposed that after the Cretaceous period was completed, in the movement which raised the mountains, of which I have spoken in § 364, there was a general elevation of the land of the northern regions, and that the severe cold which was thus produced was at least a prominent agency in the destruction of life at this period. Hugh Miller makes these two gaps or breaks the basis of a division of the life of the world into three dynasties: the dynasty of the Fish, extending from the Azoic age to the end of Palæozoic time, when the first break occurs; the dynasty of the Reptile, from this period to the end of Mesozoic time, when the second break occurs; and the last, the dynasty of the Mammal, which, commencing after the second break, continues at the present time.

366. **New Creations.**—During all the progress of the ages, whenever a new species of plant or animal appeared, there was a new creation, as already intimated in § 273. And as the life-record of the rocks shows that there have been continually extinctions of species, and introductions of new species in their places, creative power has been continually exerted in the domain of life upon

our earth. But this power was specially active when, in the great and extensive changes effected at times in the earth's crust, as was often the case, life was largely destroyed. And when the destruction was complete in the two gaps that have been mentioned, there succeeded an entirely new creation of plants and animals.

---

## CHAPTER XVIII.
### AGE OF MAMMALS.

**367. Transition from Mesozoic to Cainozoic Time.**— We now come to a period in the earth's history in which there is a decided resemblance to the present time in the general surface of the earth, and in its vegetables and animals. The surface had become diversified with numerous mountains, valleys, and rivers, though by no means as largely as now, for this diversification was completed only when man was to be ushered upon the scene. The endogens and exogens, the great sources of food for man and beast, were quite predominant in the vegetation of the earth. In the animal kingdom, the marsupials, which are allied at the same time to reptiles and to birds, gave way to mammalia of the higher orders. In the previous ages the most striking displays of animal life were mostly in the ocean, in huge fishes, and mollusks, and reptiles; but now the land was to surpass the sea in this respect.

**368. Divisions of Cainozoic Time.**—A common division of the Cainozoic age is into the Tertiary and Post-tertiary—the Tertiary occupying the time up to the Glacial period, when, as you will see, the agency of ice was applied over a large portion of the earth to produce important results, and the Post-tertiary, extending thence on to the advent of man. The term tertiary is an old geological term, which the progress of discovery has shown to be inapplicable. It was adopted when it was supposed that the crystalline rocks were the first formed in all cases, which were therefore called primary, or primitive, the secondary being those which were between the primary and tertiary in relative age. But when it was found that the so-called primary rocks were formed in various ages, some of them quite recently, this classification was given up; but the name tertiary has been retained, by common consent, as a matter of convenience. Indeed, the three terms are used still to some extent, as mentioned in § 261, primary having, however, a different meaning attached to it from what it had formerly, it being applied to the palæozoic rocks instead of those which are originally crystalline. Lyell has divided the strata of the Tertiary age into three series, according to the percentage of shells found that are the same with those that exist now. 1. Eocene, the term being derived from two Greek words, *eos*, dawn, and *kainos*, recent. When the name was adopted it was supposed that from five to ten per cent. of the species found in the strata were identical with species found at the present time, and therefore the period in which these strata were formed was very properly considered as the *dawn* of the present life of the earth. But Dana states that it has been discovered that the species of this period are all extinct. 2. Miocene, the term coming from *meion*, less, and *kainos*. Here from ten to forty per cent. of the species found are living species at the present time. 3. Pliocene, from *pleion*, more, and *kainos*. Here the percentage of living species is greater than in the miocene, it being from fifty to ninety per cent. These three series are often called **ancient** tertiary, middle tertiary, and modern tertiary.

369. **Tertiary Deposits.**—In the previous ages the strata of rocks were mostly *marine*—that is, deposited from the waters of the sea as sediment, or laid down by coral and other animals that lived in the sea. But now many of the strata were either *lacustrine*—that is, deposited

in lake bottoms; or *estuary*—that is, deposited from the waters of a branch of the sea, where there was a mingling of salt and fresh waters. Some deposits occurred in interior (mediterranean) seas. There was, during a part of the Tertiary age, such a sea in Europe of immense extent, with vast irregular islands in it. There was often in this age an alternation of marine and fresh-water deposits, as shown by the fossils found in the strata. This is the case with the basin, so called, in which the city of Paris lies. First in order we have plastic clay, a fresh-water deposit. Then upon this are strata in which there is a vast quantity of marine shells. This is the *calcaire grossier*, which is, for the most part, a granular yellowish limestone. It is regularly bedded and jointed, so that it is easy to quarry it, and it is the grand building-stone of Paris. Over 500 species of marine shells have been discovered in these strata. By some changes of the land in some way, there was at length fresh water over all that region, and in the deposits made from it upon the *calcaire grossier* there are, accordingly, found remains of fresh-water animals and plants. Again the sea water was let in, and we have a marine group of strata, to be followed again by fresh water and its characteristic deposits. In these last fresh-water beds the most peculiar rock is the millstone, a flinty rock full of cells and winding cavities, occasioned, it is supposed, by the escape of gas from the bed of the lake up through the deposit while it was hardening.

370. **Areas of the Deposits.**—The deposits of strata in this age were not made over areas of almost continental extent, as was the case in the previous periods. There were elevations of land here and there over the spaces now occupied by the continents, of such size and arrangement that there were lakes, and estuaries, and inland seas, and on the floors and shores of these the deposits were made. As these elevations multiplied and increased during the progress of the age, the areas of deposition were

less in extent in the latter than in the first part of the age. The more the areas became divided, the more diversified and complicated were the deposits.

371. **Tertiary Rocks.**—Most of the rocks of this age are less firm than those of previous ages, but some of them are very hard. They are, shell-rocks of various kinds, shell-beds, or mixtures of shells and earth, sandstones, marls, clays, beds of sand, compact limestones, conglomerates, buhrstones, etc. The lowest of the tertiary rocks in Europe were made to a large extent from materials derived from the denudation of the chalk formation. For this purpose, the cretaceous strata were upheaved from the sea in which they were formed, so that the denuding water could act upon the chalk cliffs. A slow process it was for the water to wear away sufficient amounts of these emerged strata to form the lower tertiary strata of Europe, and a long age was required to do it.

372. **Nummulitic Formation.**—I have already noticed in § 239 those small shell-animals of the tribe Foraminifera, the shells of which have formed such immense quantities of rocky strata in Europe, and Asia, and Africa. In Fig. 159, at 1 and 2, are representations of num-

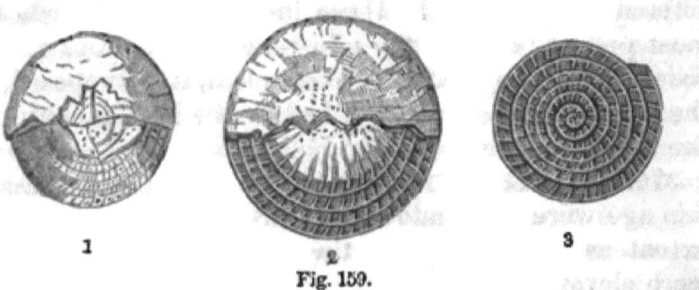

Fig. 159.

mulites as they are found in the rocks, and at 3 is a section of one, showing its cells. The resemblance of these fossils to pieces of money has given occasion to many superstitious legends in regard to them among the Germans, and they are very commonly called the devil's money. The nummulites of different beds are of different species and of various sizes, some of them reaching the size of an inch and a half in diameter. There are, of course, other shells mingled with the nummulites, though some strata are almost entirely nummulitic. "The nummulitic formation," says Lyell, "with its characteristic fossils, plays a far more conspicuous part than any other tertiary group in the solid frame-work of the earth's crust, whether in Europe, Asia, or Africa. It often attains a thickness of many thousand feet, and extends from the Alps to the Carpathians, and is in full force in the north of Africa, as, for example, in Algeria and Morocco. It has also been traced from Egypt, where it was largely quarried of old for the building of the Pyramids, into Asia Minor, and across Persia by Bagdad to the mouths of the Indus. It occurs not only in Cutch, but in the mountain ranges which separate Scinde from Persia, and which form the passes leading to Caboul; and it has been followed still farther eastward into India, as far as eastern Bengal and the frontiers of China." In the Swiss Alps nummulitic strata are found 10,000 feet above the level of the sea, and in Thibet at the height of 16,500 feet. They enter into the composition of the central and loftiest parts not only of these mountains, but of the Pyrenees, the Carpathians, and Himalayas. Where, therefore, these great mountain chains now are, the sea stood during that long age, when those little animals, the nummulites, were accumulating these vast masses of rock to be raised in after ages into lofty mountains by some of the processes noticed in § 226. Though there are some nummulites found in the strata of the same period in this country, they did not play here the magnificent *rôle* in rock-making which they did in Europe, Asia, and Africa.

373. **Tertiary Plants.**—In the earlier parts of the Tertiary period the climate of the earth was more uniformly mild than in the latter, and therefore in the strata of

that time we find in the northern portions of the earth the remains of such plants as would naturally flourish in warm climates. There are palm-like leaves and fruits, leguminous seeds of arboreal growth, twigs and leaves of mimosa, laurel, and other plants, all allied to what now grow in southern latitudes, and quite in contrast with the plants now found in the same localities. Associated with the twigs, leaves, and fruits found in the strata are beds of lignite, or brown coal, indicating a great abundance in the vegetation. From the high latitude to which warm weather extended in that period, Agassiz remarks that the Tertiary age may be called the geological summer.

374. **Strata of Diatoms.**—There are found here and there in the Tertiary system strata which are made up of what were formerly supposed to be infusorial animals. The diatoms, however, which compose by far the largest proportion of the rock, are now ascertained to be vegetable, and not animal. I have already noticed these plants, and the strata which they have formed, in § 240, and will not dwell on them here. As these diatomaceæ are so exceedingly minute, a very long age must have been required for the accumulation of thick beds of silex from their remains, such as are found about Richmond, in this country, and in Bilin, in Bohemia. As it takes 187 millions of them to make a single grain, it could have been only by the contributions of countless generations of them that a stratum of from 14 to 30 feet in thickness was laid down.

375. **Tertiary Animals.**—So far to the north was the climate warm in the Tertiary age, that as it was with plants, as stated in § 373, so it was with animals—such kinds of animals as now flourish in tropical climates were then in high northern latitudes, crocodiles, turtles, gigantic sharks, pachydermatous quadrupeds, etc. The earliest remains of birds are found in the strata of the Eocene period, the dawn of the Tertiary. So also serpents

AGE OF MAMMALS. 267

first appear in the Tertiary strata. Not a species of fish, or reptile, or bird, or mammal of the Tertiary period is in existence at the present time. These classes of animals are in contrast with mollusks in this respect, for many of the latter, as you saw in § 368, coming into existence in the Miocene and Pliocene periods of the Tertiary, are living still in the midst of the mollusks of the present period that were created long ages after them.

376. **Fishes.**—The remains of fishes abound in the tertiary strata. The most interesting are the Shark family, which flourished in great numbers both in Europe and America. Some of them were monstrous in size. In Fig. 160 is represented the tooth of the Carcharodon

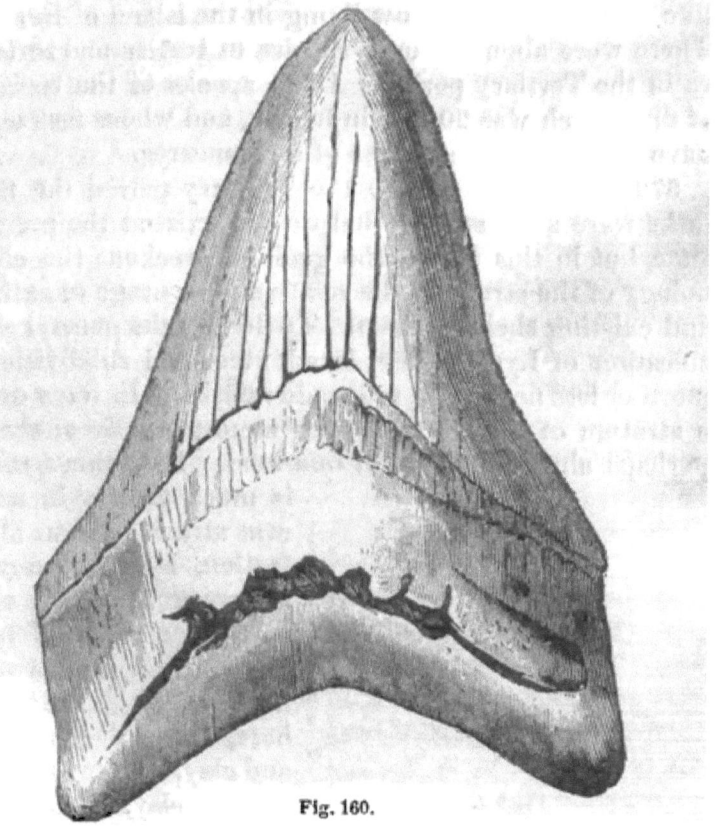

Fig. 160.

megalodon, the largest species of shark, which must have been 100 feet in length. The sharks of the present time are large, and are justly very much feared, but they are "mere pigmies," says Hitchcock, "compared with those that swam in the seas which washed the shores of North and South Carolina in the Eocene and Miocene periods." These terrific monsters, besides being much larger than the sharks of the present age, were much more numerous than now.

377. **Reptiles.**—Some eighteen or twenty species of crocodiles flourished in the Tertiary period, while there are only seven or eight at the present time. The crocodiles whose remains are found in the London clay were like those which are now living in the island of Borneo. There were about seventy species of turtles and tortoises in the Tertiary period. Dana speaks of the remains of one which was 20 feet in length, and whose feet must have been as large as those of a rhinoceros.

378. **Mollusks.**—Up to the Tertiary period the mollusks were all of species that do not exist at the present time, but in this period the geologist reckons the chronology of the strata by the relative percentage of extinct and existing shells in them. To do this the general classification of **Lyell** (§ 368) is adopted, and subdivisions, more or less numerous, are made under it. In some cases a stratum of rock is composed almost wholly of shells, perhaps almost entirely of one kind. Sometimes there is much variety in adjacent strata. As an illustration, I will give you the arrangement of a cliff on the bank of the James River, Va., represented in Fig. 161. We have here, 1. Six feet of sand and clay. 2. One foot of reddish clay. 3. A band

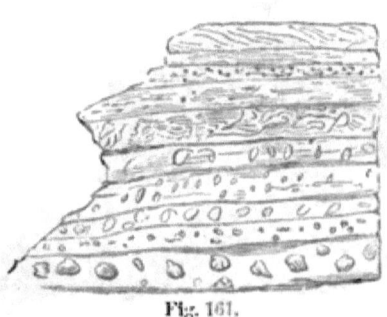

Fig. 161.

of small pebbles a few inches thick. 4. A layer of sand three feet thick, containing shells called *Chama* and *Venus difformis*. 5. A stratum four feet thick, consisting mostly of a compact mass of *Chama* and *Arca centenaria*. 6. A stratum two feet thick, mostly of large *Pectens*. 7. Another stratum of *Chama*, with *Arca centenaria, Panopea reflexa*, about six feet in depth. 8. A second layer of *Pectens*, with *Ostrea compressirostra*, one foot thick. 9. Another stratum of *Chama* three feet thick. 10. A stratum of *Pectens* and *Ostrea* five feet thick. When the shells are very perfect we know that the animals that inhabited them lived and died on the spot, and that the strata were formed very gradually in still water. We make a different inference when the shells are much broken up. On the York River, in Virginia, there is a porous rock, in some places forty feet in height, which is made up almost wholly of comminuted shells. Here was no still water, but there were the rush of the tide and the breaking of the surf during all the time that this rock was being deposited.

379. **Indusial Limestones.**—Another example, in addition to those already mentioned, of the contributions of small animals in the work of rock-making we have in the indusial limestones in the ancient province of Auvergne, in the central part of France. The agent in this case was the larva of a species of fly, allied to the common caddis-worm of the angler of the present day. We find now that the larvæ or grubs of different species use various materials for their *indusia*, or covers, some gluing together small bits of wood, others choosing grains of sand, and others still the shells of small mollusks. One of the last is seen in Fig. 162. The animal is in a tube which it has constructed for itself of minute shells, and, living in this, it thrusts its head and a portion of its body

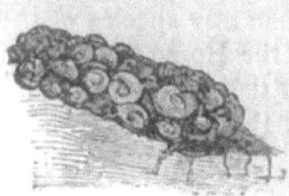

Fig. 162.

out in search of its food.* "It is evident," says Mr. Eley, an English geologist, "that the larger of the ancient lakes of Auvergne was inhabited by a species of this family, and that they swarmed in it in a remarkable manner; for their cases, incrusted with a calcareous matter, are seen to form there thick **layers of limestone**—called *indusial*, from this strange origin—which alternate with the more usual kinds of marl through a thickness of several hundred feet." These indusial strata are **eight or ten feet** thick, and extend over an area of many square miles. When you are told that one of the cases or tubes contains over a hundred shells, and ten or twelve tubes may be counted in a single cubic inch of rock, you may have some idea of the countless myriads of minute mollusks which must have formerly lived and died in every part of this region, and of the length of time required for the formation of the strata constructed chiefly from their shells.

380. **Mammals.**—Animals of this class were specially prominent in the scenes of this age. Tapir-like pachyderms flourished largely in the early part of the age, but as we come to its latter portions in the examination of the strata, we find the mammals more like those of the present age. Still, there is not a single species of all those that have this resemblance which is not now extinct. There are some localities that are peculiarly rich in remains of the mammals of the Tertiary. In the Upper Missouri region, in this country, among other remains, there have been found those of three species of camel, a rhinoceros as large as the Indian species, a mastodon, an elephant, a wolf larger than any species of the present day, four or five species of the Horse family, etc. It seems strange to us that some of these animals should exist in such a locality, so far from the haunts of similar animals in the present period; but more strange still is

---

* For a full description of the habits of these animals, see my Natural History, § 459.

it to think of monkeys as being in England, and yet remains of them are found in that locality as far back even as the strata of the Eocene.

381. **Cetacea.**—The most remarkable of the **Cetaceans** of this age is the Zeuglodon cetoides, a **tooth of which** is seen represented of its natural size in **Fig. 163.** The

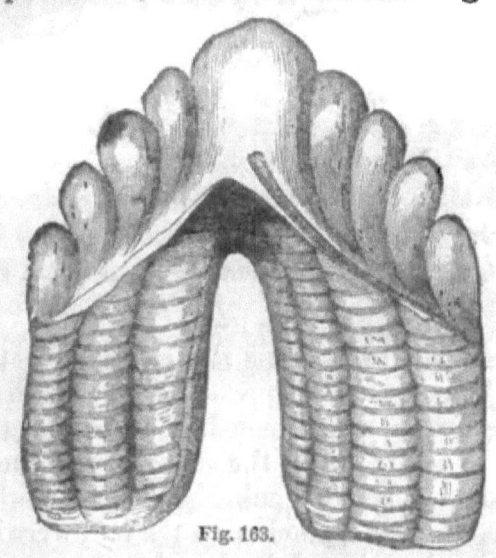

Fig. 163.

tooth is yoke-shaped, and hence the name given to the animal, which is derived from two Greek words, *zeuglon*, a yoke, and *odons*, tooth. Its remains were first noticed by Dr. Harlan, of Philadelphia, who in 1832 described a vertebra weighing 44 pounds, which was brought from the Washita. He referred the animal to which it belonged to a new genus, which he called Basilosaurus—that is, king of the Saurians, supposing it to be reptilian in character. It has since, however, been proved to belong to the whale tribe. Its vertebræ were formerly so abundant in the State of Alabama that they were used very commonly in making walls. Some of them are a foot and a half in length, and a foot in diameter. Lyell saw the vertebral column of one skeleton which extended on the ground nearly 70 feet.

**382. Pachydermata.**—The most prominent of the mammals of the early Tertiary were the pachydermatous quadrupeds. Some of these animals are represented in Fig. 164. The one on the right is the Palæotherium.

Fig. 164.

This is intermediate in shape and character between the tapir and rhinoceros of the present day. There are about twelve species, varying much in size, the largest being of the size of a rhinoceros, the smallest about as large as a sheep. The other two animals, the Anoplotherium in the middle, and the Ziphodon on the left, were lighter in their construction, though having the same general character with the Palæotherium. These are representatives of a large class of mammals that flourished in that age in the forests, on the plains, and by the lake and river-swamps. They were curious creatures, uniting in themselves the peculiarities of two or more animals of different, even often of opposite character, such as tapir, sea-cow, hog, rhinoceros, ass, camel, and antelope.

**383. Cuvier's Investigations.**—The remains of these pachydermata have been found in great abundance in the strata of what is called the Paris basin. Four fifths of the fifty species of quadrupeds whose remains are found there belong to this tribe, which is represented in the present age by only four species, so different are the prevalent forms of animal life now from what they were in the Eocene of the Tertiary. Cuvier was the first to develop the true character of the remains in the Paris basin. Soon after he had investigated the fossil elephants (to be noticed hereafter), some bones found in the quarries of Montmartre, near Paris, were brought to him for examination. He at once instituted comparisons between these and the bones of various animals of the present age, in order to ascertain the character of the animals to which the fossil bones belonged. He soon decided that they were pachyderms; and at length, with his knowledge of the relations of bones to each other and to other parts of the frame, he was able, from a few bones, such as those of the head, the jaws, some teeth, and a few of the vertebræ, to fill out the deficiencies, and he drew the outlines of two animals as he supposed them to have been, the Palæotherium and Anoplotherium, represented in Fig. 164. He presented the subject to the French Academy, averring that the fossil bones belonged to animals of a creation previous to the present animal creation, and he therefore called the larger animal Palæotherium, from two Greek words, *palaios*, ancient, and *therion*, animal. These views were received variously by scientific men, some disbelieving, many doubting, and those only believing who had paid some attention to comparative anatomy. As the investigation proceeded, complete skeletons of the animals were after a while found in the strata, and the outlines made by Cuvier were shown to be correct, and his views were universally adopted.

384. **Dinotherium.**—Farther along in the Tertiary than the pachyderms of the Paris basin appeared a monstrous pachyderm of a singular composite character, called the Dinotherium, represented in Fig. 165 (p. 274). Its remains have been found in various parts of Europe, the most perfect being a fine skull found in Germany. The figure gives the animal as it was restored by Professor Kaup from the scanty remains which have been obtained. As none of the bones of the limbs have as yet

Fig. 165.

been found, there is some difference of opinion among naturalists as to the true character of the animal, some supposing it to be much like an elephant, and others allying it to the dugong, a swimming pachyderm, for the description of which I refer you to my Natural History, § 195. There is, in truth, in this strange animal, a mixture of the peculiarities of the elephant, hippopotamus, tapir, and dugong. If it be a quadruped, it is the largest of all the quadrupeds that have lived on the globe, being larger than even the mammoth and mastodon, hereafter to be noticed. It was probably eighteen feet long. Its skull is nearly four feet in length, its upper jaw being shaped like that of the elephant, for the attachment of a trunk. It had two enormous tusks on the lower jaw, curving downward like those of the walrus. It probably lived chiefly in the water, like the hippopotamus. Its diet, as we know by its teeth, was vegetable, and it undoubtedly used its tusks to tear up the roots of aquatic plants by a rake-like action from the beds of rivers and lakes. Perhaps, too, it used its tusks as hooks for drawing its huge unwieldy body up banks, and even along upon the ground if it had no real legs.

385. **Tertiary Mountain-making.**—There was, as you have already seen, considerable mountain-making during the Tertiary period, and therefore much change of elevation and flexure of the strata, with more or less of frac-

tures. The Rocky Mountains, which began to be raised at the close of the Cretaceous age, did not reach their full height till late in the Tertiary period. The Pyrenees and Carpathians were lifted up in the Eocene part of the Tertiary. It was during the Tertiary age that the Alps, Appenines, the Caucasian range, and the Himalayas attained, very nearly at least, to their present altitude. On the completion of this period the mountains of the earth were very generally raised to their full height.

Observe that these mountains, which were raised thus late, comparatively, in the course of the formation of the continents, are many of them of very great height. Compare them with the Alleghanies, that were raised, as is supposed, at the close of the Carboniferous age. The contrast is still greater if you compare them with the Laurentian Hills, so called, in Canada, which were the mountains in the Azoic age on that long island, the germ of the North American continent, spoken of in § 267. These first mountains of the earth are nowhere more than 1500 or 2000 feet above the level of the sea. The reason of this difference between the older and more recently formed mountains is thus given by Agassiz in speaking of the Laurentian Hills. "Their low stature, as compared with that of more lofty mountain ranges, is in accordance with an invariable rule, by which the relative age of mountains may be estimated. The oldest mountains are the lowest, while the younger and more recent ones tower above their elders, and are usually more torn and dislocated also. This is easily understood when we remember that all mountains and mountain chains are the result of upheavals, and that the violence of the outbreak must have been in proportion to the strength of the resistance. When the crust of the earth was so thin that the heated masses within easily broke through it, they were not thrown to so great a height, and formed comparatively low elevations, such as the Canadian hills, or the mountains of Bretagne and Wales.

But in later times, when young, vigorous giants, such as the Alps, the Himalayas, or, later still, the Rocky Mountains, forced their way out from their fiery prison-house, the crust of the earth was much thicker, and fearful indeed must have been the convulsions which attended their exit."

With the raising of so many mountains, the great systems of rivers alluded to in § 165 were begun; but, as they were not completed till the next age, I will defer the consideration of this subject till I come to speak of that age.

386. **Tertiary Volcanoes.** — There are evidences that volcanic agencies were at work during the Tertiary age in various parts of the earth. Such evidences have been found in the West India islands, in Central America, in parts of the Andes, and on the eastern continent in France, Italy, Spain, Greece, and the island of Sicily. The most interesting and striking evidences are found in the ancient provinces of Auvergne, Velais, and Vivarais, in the centre of France. Many hundreds of cones are seen there, and the streams of lava which issued from them can be traced, in many cases, as distinctly as those which in this present age so recently have flowed from Vesuvius or Etna. Some of these cones are represented in Fig. 166. Many of them have been preserved with

Fig. 166.

remarkable distinctness. This is owing to their loose, porous structure, which at first thought would be considered as rendering them very destructible. The explanation is, that all the rain which falls is absorbed at once by the porous mass, so that there are no rills running down the sides of the cones to wear them away. The highest of the volcanoes is Mont Dor, in Auvergne, which was several thousand feet above the surrounding platform, and retains now the shape of a flattened cone, broken on its summit into several rocky peaks, probably from occasional earthquakes in ages long gone by, and the continued influence of rain through not merely many centuries, but many ages.

387. **Basins.**—From the disturbances which occurred in the Tertiary period, there are strata found in some localities so arranged in flexures as to form what are called basins. Among the most noted of these are the basins of London and Paris. In Fig. 167 is represented, in a

Fig. 167.

rude way, that of London. At 4 we have the chalk lying upon green sand, 5. Upon the chalk lies the *plastic* clay, so called because in France, where it is less mixed with other materials, it is extensively used in pottery. Here there are in it beds of flint and pebbles alternating with sands and clay. Upon it lies the London clay, 2, upon which the city of London stands. It varies from two to six hundred feet in thickness, is of a dark color, tough in texture, and having mixed with it here and there earth of a green color, white sand, and masses of clayey limestone. At 6 is the River Thames. At 1, 1 are caps of marine sand, which are found on many of the hills in the valley of the Thames. This sand probably, when first deposited, formed a continuous bed over the region, but has been excavated with a part of the London clay, an example of the denuding agency of water.

With such a basin arrangement of strata as we have here and at Paris, you can see how artesian wells can be successful in both of these localities, as noticed in Part I., § 120, to which I refer you for the explanation of their operation.

**388. Tertiary Continent-making.**—In the Tertiary age a great work was done in the building up of the continents. But it was not so much by the laying down of deposits, though these were vast in thickness and over extensive areas, as it was by elevating strata already made, and crumpling them up in various quarters into mountains. The work of this period is thus summed up by Dana. "There was, 1, the finishing of the rocky substratum of the continents; 2, the expansion of the continental areas to their full limits, or their permanent recovery from the waters of the ocean; 3, the elevation of many of the great mountains of the globe, or considerable portions of them, through a large part of their height." He says, also, that "the mass of the earth above the ocean's level was increased two or three fold between the beginning and end of the Tertiary period." In this country there was a mere fringe added along its eastern and southern edge, but in the west, beyond the Mississippi, there was an addition of vast extent. The North American continent, with its triangular shape, had now its great triangular skeleton of mountain ranges, as Agassiz calls it, completed, shutting in that immense area, the Mississippi Valley. The Laurentian Hills on the north, making the base of the triangle, came into being in that first long, dark age of the world, the Azoic age; the range of Alleghanies, making the eastern limb, were raised probably at the close of the Carboniferous period; and the Rocky Mountains, the western limb, were thrust up in the Tertiary age. The continent was now essentially completed, the only great works yet to be done being the filling up of a narrow gulf which extended to where the city of St. Louis now is, from the material brought down from wide regions by the Mississippi and Missouri rivers, and the building up of the peninsula of Florida by the coral animals, which are at the present time building as busily as ever.

389. **Post-tertiary Period.**—This period extends from the conclusion of the Tertiary to the advent of man. The building of all the continents was now, as I have just said of North America, essentially completed—that is, the land had now reached its present bounds, and its grand ranges of prominences were raised to their destined heights. But there needed to be a greater diversification of surface than had as yet been effected—a diversification for the most part in detail. Hills and valleys were to be made; stones of various sizes were to be scattered over the surface; rivers were to be extended and multiplied; and terrace-like formations were to be made along rivers, and lakes, and seas. Moreover, it was necessary that much rock should be broken and ground up, to make a sufficient quantity of fertile earth for man and the accompanying races of animals. All this was done by the agency of water in its two forms, liquid and solid, as you will see as I proceed.

390. **Divisions.**—This period has been variously divided. As the system is developed in this country, it is divided by Dana into three epochs. First is the *Glacial*, when, in the higher latitudes, the land was raised much above its present elevation, and arctic cold prevailed over a large portion of the earth from either pole toward the equator, producing glaciers and icebergs. The second is the *Champlain*, so named because the beds of this epoch are well developed at Lake Champlain. In this epoch there was, in strong contrast with the Glacial, a depression of the surface below its present level, and there were deposits of three kinds—river-border, lacustrine, and marine, or sea-border. As the land sank down, the glaciers and icebergs melted, and the waters swept over the varied surface of the land, producing

great changes in the disposition of its loose materials. The third epoch was the *Terrace* period. Now there was gradual elevation of the land till it reached its present level, the waters all the time acting upon the materials laid down in the Champlain epoch, especially in the rivers. This action made terraces skirting rivers, lakes, bays, etc., which fact has given the epoch its name. It is a transition epoch, for it is separated by no well-defined line from the succeeding age, the age of Man.

The same division can be made essentially in Europe. I will speak of each epoch separately.

391. **Glacial Epoch.**—In this epoch, the summer heat, which prevailed over the now temperate regions during the Tertiary age, was succeeded by arctic cold. The cap of ice which now covers each pole extended then far toward the equator. In this country, all New England, New York, and other parts in the same latitudes were covered by it. In Europe, those parts that have now glaciers far up in the valleys of their mountains were then covered with them as Greenland is now. Ice reigned then all over Great Britain. The change in temperature was a gradual one, produced, as it is supposed, by the elevation of all the land in a body at the north to a much higher level than it has at the present time. The glaciers of that period moved down in the valleys as the glaciers of the present time do. One moved down the valley of the Connecticut, another down the valley of the Hudson, etc. They produced results similar to those which are produced by glaciers at the present day, described in § 190. Many of them were thicker and larger than present glaciers, and produced, therefore, larger results. There were then, as now, icebergs wherever the glaciers reached to water instead of terminating on land.

392. **Drift.**—As a consequence of the action of glaciers and icebergs in the Glacial period, there is scattered over all the northern parts of America, Europe, and Asia what

is called *drift*. It is various in its composition, the material being sand or gravel, or boulders of various sizes. Sometimes the boulders are mingled with the sand and gravel, and sometimes they are separate. When the materials of drift are in manifest layers, and to some extent sorted, the finer and coarser alternating, and the fragments are rounded and smoothed, it is called *modified drift*. Water has in this case acted upon the materials, and laid them in beds. Now of all this material which we call drift, none was produced where it lies, but it was transported to its localities, and for the most part from the north toward the south. There are two theories in regard to its transportation—the one attributing the result to glaciers, and the other to icebergs—called respectively the glacier and the iceberg theory. But the facts show conclusively that neither could alone accomplish all the work, and that both must have been more or less in operation. The drift must have produced great changes on the surface, filling up valleys here and there, making lakes to overflow, and altering the courses of rivers. A marked instance we have of the latter change in the case of the Niagara River. There is decisive evidence that the bed which it flowed in, from the whirlpool onward, until the Glacial epoch was then filled up with drift, and the water opened for itself a new gorge through solid rock, through which it has run to the present time. The drift is by no means confined to the valleys and plains, but extends high up on the sides of mountains, and is found even on the tops of some of them, as Mount Holyoke and Mount Tom, in Massachusetts. It is at the height of 2000 feet in the Green Mountains, 3000 on Monadnock, and even 6000 on Mount Washington.

393. **Boulders.**—These are angular, or more or less rounded, according to the amount of friction to which they have been subjected by water and other stones. They vary much in size, and while they commonly do

not exceed a cubic foot, they sometimes reach thousands of cubic feet in size, and thousands of tons in weight. The boulder on which the colossal statue of Peter the Great stands in St. Petersburg is a mass of granite weighing 1500 tons. A boulder at Whittingham, Vt., a town in the Green Mountains, is 43 feet long and 32 in average width, contains 40,000 cubic feet, and weighs 3400 tons. The distances to which boulders have been transported have been much investigated. The ordinary distances are from 20 to 40 miles, but they have often been carried 60, or even 100 miles. Hitchcock speaks of some boulders found in Ohio and Michigan which came from the ancient azoic rocks of Canada, and calculates that they must have been brought from a distance of from 400 to 600 miles. These distances are discovered by comparing the boulders with the rocks of the country, thus tracing them back to the sources from which they came. Sometimes great boulders have been carried across deep valleys. Thus the monstrous one in Whittingham was transported across a valley 1000 feet deep, and another on the Hoosac Mountain, in Massachusetts, came across a valley 1300 feet deep.

Sometimes many square miles are almost covered with boulders of various sizes. This is often seen in the hilly parts of New England. A singular circumstance in regard to the disposition of boulders, first pointed out by Hitchcock, is the arrangement of them in long trains. There are two nearly parallel trains in Massachusetts, extending from between Canaan and Lebanon, one of them to a distance of 20 miles and the other 10. They are from 300 to 400 feet in width, and are about half a mile apart. It is a singular fact that they cross two ranges of mountains which are 100 feet higher than the source from which the stones came. The boulders are most of them large, one of them weighing over 2000 tons, and their angles are but little rounded, showing that they have been subjected to but little friction. Did two enormous icebergs float side by side slowly over that region, dropping their freight of stones as they melted away? This is the only supposition we can make, but this is not at all satisfactory.

394. **Glacial Markings.**—In speaking of the effects of glaciers in § 190, I referred to various markings made by them on the rocks of their beds. Similar markings are found on the sides of mountains, and on the faces of rocks where the glaciers and icebergs of the Post-tertiary age moved along. We find them every where in the tracks of boulders. There is great variety in them. Sometimes there are deep furrows, and sometimes mere scratches. Sometimes there are lines as delicate as if made by the tool of an engraver, and sometimes the surface of a stone susceptible of polish is as smooth as it would be if it were artificially polished in the shop of a manufacturer. Sometimes rocky ridges have been rounded and smoothed, giving them an undulating appearance, like the *roches montonneés*, or embossed rocks seen in the Alps, the result of the action of glaciers of the present time, as represented in Fig. 42. It is evident that most of these markings must have been made by glaciers, and not by icebergs. While icebergs can plow up dirt, and sand, and gravel, and crush rocks, it requires the firm, slow, steady movement of the glacier, holding its rocky tools imbedded in the ice, to make the regular parallel grooves and lines, and the polished surfaces. And, in proof of the correctness of this view, we find a striking resemblance between the effects of modern glaciers on the rocks and those marks which appear where the glaciers of the Post-tertiary age were supposed to be. Still there is some coarse and irregular work, as we may term it, which must have been done by icebergs, as, for example, the making of furrows and valleys on the summits of some of the lesser mountains.

The glacial markings are found abundantly on high mountains as well as in valleys and on plains. On the

Green Mountains they are found at the height of 6000 feet.

395. **Champlain Epoch.**—As the land elevated in the Glacial epoch subsided, and water rose over it, the Champlain epoch was ushered in. It found a vast quantity of material which had been broken up by the ice of glaciers and icebergs, and transported by them from the sources from which it was obtained, mostly in southern directions, and to greater or less distances. The finer portions of this, the gravel, and sand, and mud, the moving waters of this epoch carried about and laid down in widely-extended strata. These formations are *alluvial*, the term being derived from the Latin word *alluo*, to wash. Boulders never are found lying upon these beds, for they were scattered by the glaciers and icebergs before these strata were laid down. They are often enveloped and covered up by the strata. They are often also found on elevations in the neighborhood of Champlain strata. In such cases there may have been once some of the finer material of the drift mingled with them, and this may have been washed away, the water not being able to remove the boulders themselves. As the land subsided or sunk down extensively in this epoch, some of the deposits occurred on very high elevations. Hitchcock mentions beaches of these deposits at great heights in the Green and the White Mountains, one at the Franconia Notch being 2665 feet above the level of the sea. The subsidence of the land at the north during this epoch, lessening the amount of it above the surface of the water, and so removing the arctic cold, which the raising of the land in the Glacial period had produced (§ 391), melted, of course, the glaciers of the temperate regions. This created at the outset of this epoch a great flow of waters over the continents, which carried with them large quantities of the smaller materials of the drift. The deposition of these was much modified by the lakes and rivers which already existed, and, in return, also made many

changes in them. It is difficult, in many cases, to distinguish the deposits of this period from those of the succeeding one, but it may be remarked in the general that the river-plains and sea-beaches of the Champlain epoch are at the present day *elevated* plains and beaches—that is, existing at higher levels than the terraces made in the succeeding period.

396. **Terrace Epoch.** — As you have seen, there have been, in the course of the formation of the continents, many extensive subsidences of land, letting the water prevail for a time where it had been dry land before. The last of these general submergences occurred in the Champlain epoch. In the next epoch, the Terrace period, the movement of the land was *upward*—a movement of emergence—which went on until the land acquired a comparatively settled condition—that is, one from which it has not varied, to any great degree, since the advent of man. During this upward movement of the Terrace epoch there was, to a considerable extent, a rearrangement of the strata laid down in the Champlain epoch. In this rearrangement the formation of terraces was so common that this has given the name to the period. The term *terrace* is applied to banks of loose materials skirting the sides of valleys about rivers and lakes, having a level surface on the top, and fronting on the river or lake with a more or less steep escarpment. As they are often quite numerous, rising above each other like the seats of an amphitheatre, but differing much in width and in the lines of their edges, they give great variety and beauty to the scenery, especially when the habitations of men are built upon them, which is often done, or when, as is sometimes the case, they are used as the dwelling-places of the dead. Not only do they vary in width and in line of edge, but the variety is often increased by the action of rain with the currents caused by it on their level-topped surfaces. You see in Figs. 168 and 169 (p. 286) sections of two series of ter-

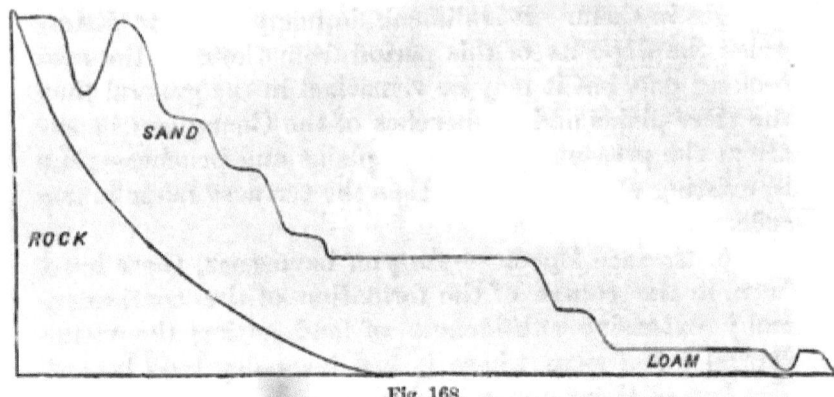

Fig. 168.

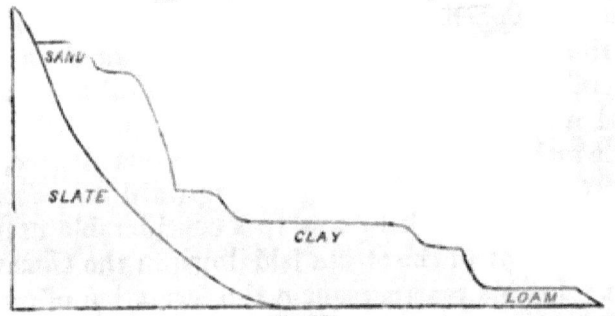

Fig. 169.

races. In Fig. 168, at the right hand, is represented a section of the bed of the Connecticut River, lying in the alluvial loam. Here there are seven terraces of various widths, the upper one having a considerable excavation in it from the continued action of rain. In Fig. 169 is represented a set of terraces at another locality on the Connecticut, where the arrangement is somewhat different.

397. **How Terraces were Formed.**—Terraces resulted from the action of water on the modified drift as the land was rising from its subsided condition in the Champlain epoch. In other words, they were formed by the natural drainage of the country during that rise. Of course, various circumstances had a play in their formation, giving them variety in their shape and arrangement. The

operation of the chief of these circumstances is stated by Dana after this manner. Every river has its channel and its flood-plain, the latter being overflowed whenever a freshet occurs, as stated in § 174. Now if, when the land was rising, the rise was chiefly in the interior of a continent, the coast being little changed, the rivers running toward the sea would have their slope increased, and so greater force would be given to the descending water. The result of this would be, that not only would the channel be deepened, but a part of the flood-plain would be at a lower level. This is indicated in Fig. 170,

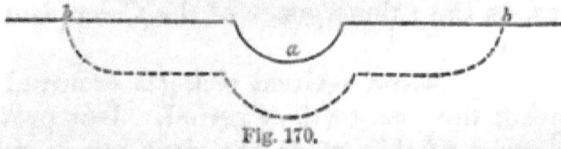

Fig. 170.

where $a$ is the bed of the river, and $b$ the flood-plain, in the Champlain period. The change thus effected by the erosive power of water is represented by the dotted line. This would leave the river as seen in Fig. 171, with $b$ for

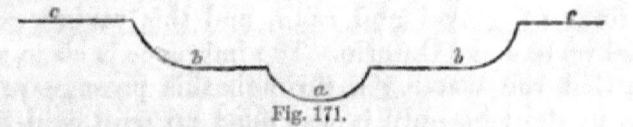

Fig. 171.

the flood-plain, and a terrace, $c$. As soon as the river by this process attained its former slope, this process would stop. Then, if the same rising went on, another similar result would occur. And so with a succession of risings, or perhaps with one continued rising, several terraces might be made, as represented in Figs. 168 and 169. There are many circumstances which vary the results of this process, and make the terraces irregular in height and form. There are also other processes that have produced terraces, but these I will not stop to describe.

398. **Sea-beaches.**—I have spoken of the beaches which were made in the Champlain epoch. When these are found in the neighborhood of terraces they lie at a high-

er level, though, as they were formed earlier than the terraces, they extend down under them, and are therefore, geologically speaking, lower. They often form a sort of upper terrace in the series, making a fringe along the sides of the hills that shut in a valley. Though irregular in form, they have a certain general level, if observation be made of one for any considerable length along the hills that it skirts. They consist mostly of sand and gravel, all the stones in them being rounded, thus showing the influence of long-continued water friction. It is supposed that they are beaches left by the retiring sea, as the **submergence of the Champlain period** passed away.

399. **Niagara River.**—Great changes occurred in this river during the Post-tertiary period. Just previous to the beginning of this epoch the river ran in entirely a different bed from what it now runs in from the whirlpool on to its outlet in Lake Ontario. The evidence is this: The west bank of the gorge at the whirlpool shows the beginning of a deep ravine filled with drift in the form of gravel and sand, and this ravine can be traced on to Lake Ontario. The inference is clear, therefore, that the water ran through this passage of four miles to the lake until it was filled up with drift in the Glacial period; and this bed being filled up, the water sought, and, to a great extent at least, made for itself, by its erosive power, another passage, the one in which it now runs. But this is not all. The water has effected in this passage, as I have before stated in § 183, the removal of the falls from the neighborhood of Queenstown back seven miles to their present position. The process by which this has been done has been explained in § 183; but the story of the changes which have been effected in the Niagara River is not all told yet. The still waters of a lake once lay over all the neighborhood where the falls now are, for there are lake deposits there, as shown by the fossils which the strata contain. The

tooth and other bones of a mastodon found in these deposits show that they were laid down in the Champlain epoch, for this animal flourished at that time.

**400. Length of the Post-tertiary Age.**—If we can not make some approximation to an estimate of the length of time consumed by the Post-tertiary period, by calculations from the rate of progress in the recession of the Falls of Niagara, we can, at least, see that the age was a very long one. The recession is still going on, and various observations have been made in regard to it, and estimates have been based on those observations. The results, as worked out by different observers, vary greatly. Some estimate the recession as high as three feet in a year, which undoubtedly is far beyond the fact. Lyell estimates it as averaging one foot a year. This, which Dana says is certainly large, would give over 31,000 years for the whole recession. Mr. Desor, another observer, inferred from the data which he ascertained that it was "more nearly three feet a century than three feet a year." Taking it at three feet a century, the whole would require over a million of years.

**401. Post-tertiary Rivers.**—Rivers are a part of the grand apparatus for the circulation of water on the earth. Evaporation carries up water in the atmosphere from over the whole surface, from the land as well as the sea, and the vapor thus dissolved in the air is condensed into rain, or snow, or hail, in order to be brought back to the earth. Falling, it gathers in little streams, which, uniting together, make the great rivers that run to the ocean. Now, in order to have very large rivers, two things are necessary—a considerable extent of land to afford room for the union of many streams in one, and the presence of great mountains, to condense the vapor rapidly with their tall, cold peaks. Neither of these conditions was present when the continents began to form. When that long island, with its low mountains, the germ of the North American continent, was lifted up in the Azoic age from amid the waters, no such rivers as the Mississippi or the St. Lawrence were upon it. And the grand river systems of the earth were not fully formed till after the continents were expanded to their full limits, and the mountain ranges spoken of in § 385 were raised up in the Tertiary age. The Post-tertiary age then was, as stated by Professor Dana, "*the era of the first grand display of completed river systems*—of the first Amazon, Mississippi, Ganges, Indus, Nile," etc.

**402. Terminal Moraines of the Ancient Glaciers.**—The glaciers of the Post-tertiary age had terminal moraines, like those of the present time (§ 190). They have, of course, been much altered by the fluviatile operations, that is, the operations of moving water, which produced such great effects during the long ages of the Champlain and Terrace periods, and hence it is that I notice them here. The alterations alluded to are so great that the resemblance of the remains of these ancient moraines to those found at the foot of the glaciers of the present day is not obvious to the common observer. It is only the practiced eye of the skillful geologist that can discover it. Agassiz, Guyot, and others have investigated this subject by extended observations in some of the localities where the vast glaciers of the Glacial period lay, especially in Switzerland and the island of Great Britain, and have traced out the moraines with signal success, in spite of the obliteration of their distinctive marks by causes which have been acting upon them for ages. They are, for the most part, semicircular in shape, the concavity being toward the direction from which the glacier moved downward. Sometimes there are several of these semicircular walls, one placed within another. Each of these is a moraine, the outer one having been formed first, and the others one after another, as the glacier retreated or became shortened (§ 189). In the passing away of the Glacial age there was, of course, a constant diminution of the extent of the glaciers, and consequently many concentric moraines were formed. I have spoken of the moraines as walls, but many of them are broad and extensive, some of them being now the sites of cities. The cities of Berne and Zurich stand on moraines. Agassiz mentions a moraine through which a river had made for itself a passage, and a village occupies the moraine on both sides of the river.

These ancient moraines, like those found at the present time, are made up of boulders, pebbles, gravel, and sand, indiscriminately mixed together. But their composition is concealed from view by the soil which has accumulated upon their surface in the many centuries, even long ages, that have passed since their formation, and the consequent vegetation that has sprung up upon it. "Time," says Agassiz, "which mellows and softens all the wrecks of the past, has clothed them with turf, grassed them over, planted them with trees, sown his seed, and gathered in his harvests upon them, until at last they make a part of the undulating surface of the country."

403. **Soil made in the Post-tertiary Period.** — Soil is merely comminuted rock. Nothing but the lowest order of vegetation can grow on solid rock, and that in the scantiest manner. The rock must be broken and ground up into soil in order to be the basis of a full vegetation. The various agencies by which this is done have been brought to your view in various parts of this book. They have been at work at all times since the first rocks of the Azoic age were formed, but at some periods more than at others. They were especially at work in the Post-tertiary period. Quite a large portion of the soil now cultivated was produced then, and we may say that it was one of the great objects of the Creator, in the operations of that period, to provide soil for the gardens and fields of man, who was to come upon the scene of action in the following age. The rocks were broken and ground up for this purpose by the glaciers and icebergs of the Glacial period, and the work of grinding and sorting was

continued on through the Champlain and Terrace periods, the water making piece rub against piece, great and small, thus furnishing the small particles that were to be the nutritious part of the soil which it was to spread over the country. Hugh Miller, in speaking of the soil which was thus furnished to Scotland, holds the following language: "It is but a tedious process through which the minute lichen, or dwarfish moss, settling on a surface of naked stone, forms, in the course of ages, a soil for plants of greater bulk and a higher order; and had Scotland been left to the exclusive operation of this slow agent, it would be still a rocky desert, with perhaps here and there a strip of alluvial meadow by the side of a stream, and here and there an insulated patch of mossy soil among the hollows of the crags; but, though it might possess its few gardens for the spade, it would have no fields for the plow. We owe our arable land to that geologic agent which, grinding down, as in a mill, the upper layers of the surface rocks of the kingdom, and then spreading over the eroded strata their own *débris*, formed the general basis in which the first vegetation took root, and in the course of years composed the vegetable mould. A foundering land under a severe sky, beaten by tempests and lashed by tides, with glaciers half choking up its cheerless valleys, and with countless icebergs brushing its coasts and grating over its shallows, would have seemed a melancholy and hopeless object to human eye had there been human eyes to look upon it at the time; and yet such seem to have been the circumstances in which our country was placed by Him who, to "perform his wonders,"

"Plants his footsteps in the sea,
And rides upon the storm,"

in order that, at the appointed period, it might, according to the poet, be a land

"Made blithe by plow and harrow."

**404. Post-tertiary Animals.** — There are three points

worthy of remark in relation to the animals of this period. 1. Their great size. The elephants, bears, lions, horses, hyenas, etc., were much larger than the species of these animals existing at the present time. Perhaps the difference is greater in the sloths than in any other class of animals, as you will soon see. 2. The division as to the general character of prevalent species in the different continents was much the same as now. Thus, in the eastern continent, the Carnivorous animals predominated; in North America, the Herbivorous; in South America, the Edentates; and in Australia, the Marsupials, or pouched animals. 3. Animals that were in character like those which now have their habitat in warm climates lived then in the regions that are now temperate, and in some cases even up to the arctic regions, as in the case of the elephant and rhinoceros of Siberia. Owen says of England at this period, "Gigantic elephants of nearly twice the bulk of the largest individuals that now exist in Ceylon and Africa roamed here in herds, if we may judge from the abundance of their remains. Two-horned rhinoceroses of at least two species forced their way through the ancient forests, or wallowed in the swamps. The lakes and rivers were tenanted with hippopotamuses as bulky, and with as formidable tusks as those of Africa." Besides these, there were tigers larger than those of Bengal, and "troops of hyenas larger than the fierce *Hyena crocuta* of South Africa, which they most resembled, crunched the bones of carcasses relinquished by the nobler beasts of prey, and doubtless themselves often waged a war of extermination on the feebler quadrupeds."

I will now go on to notice some of the animals that were peculiar to this period.

405. **Post-tertiary Elephants.**—*Elephas primigenus*, the great Siberian mammoth, was a third taller than the largest of elephants of the present time, and was twice as heavy. The skeleton of this animal is represented

in Fig. 172, and the animal itself is seen in Fig. 173,

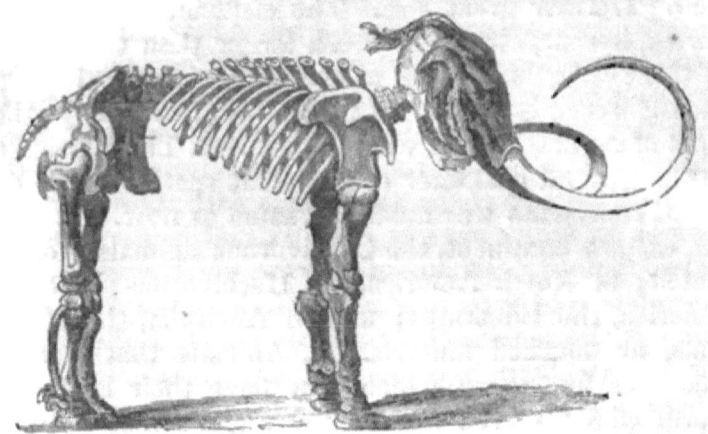

Fig. 172.

Fig. 173.

at the left side. You observe that the tusks are very

much longer than in elephants of our time, and that they are curved upward and backward with a broad sweep. This animal had long **black hair,** mingled with which there was a coat of reddish-brown wool. Such a covering undoubtedly fitted it to bear a colder climate than it could bear with the ordinary covering of elephants. At the same time, it is clear, from the fact that so many animals of the same kinds with those that now flourish in **warm regions** then existed in northern latitudes, that the climate of the far north was, at least, much less cold than it is now. Indeed, the woolly covering of the mammoth of Siberia simply indicates that the climate there was only so much cooler than it is where elephants are accustomed to live now, that this additional covering was required for its comfort, and yet it was not so cold as absolutely to forbid the existence of animals of that nature, as is the case in that region at the present time. Besides, if **arctic cold reigned there** as it does now, the vegetation **could not have been** sufficient for the sustenance of any number of these enormous quadrupeds. The remains of the mammoth show that great herds lived there. The tusks which are found furnish a large part of the ivory in the market. The Liakhow Islands, lying off the north coast of Asia, are composed to a great extent of mammoth bones, cemented together by sand and ice, and in the year 1821 as much as 20,000 pounds of fossil ivory was obtained from the island of New Siberia, some of the tusks weighing nearly 500 pounds. The mammoth lived in herds in England. Its remains have also been found in North America, but not much below the latitude of 40°. Below this level in this country another species of elephant flourished, and was very abundant in the South, in the Valley of the Mississippi. On the island of Malta there have been found the bones of a pigmy elephant which was about the size of a calf.

406. **The St. Petersburg Skeleton.**—There is a skele-

ton of a mammoth in the Imperial Museum at St. Petersburg, the story of which is interesting, both on account of the circumstances under which it was obtained, and the influence which it had upon palæontological researches. The story is this: In the year 1799 a Siberian fisherman saw a rounded mass projecting from an ice-bank near the mouth of the River Lena. The summer weather so thawed it year after year that in 1803 the enveloping ice was all melted, and the nucleus of this mound-like projection was found to be an enormous elephant. Though it had been there not merely centuries, but ages, it was perfectly preserved, so that dogs and wolves fed upon it as upon fresh meat. In the next year, 1804, the fisherman cut off the tusks, which weighed 360 pounds, and sold them. In 1807 an English traveler, Mr. Adams, hearing the story, visited the spot, succeeded in collecting all the bones except those of one foot, which were supposed to have been carried off by wolves, and recovered also the tusks. He also found some of the hair and wool, and parts of the skin. The name Mammoth, which was given to this elephant, is a Siberian word meaning earth-beast, the idea of the natives being that the mammoths live somewhere under ground, and die whenever they come to the surface and feel the influence of the sun.

407. **Cuvier's Views.**—The discovery of the Siberian elephant was the occasion of many speculations. At first it was thought that it was transported to this high latitude from India by some accident, and similar remains found shortly after in Italy, Germany, etc., were supposed to belong to Carthaginian elephants brought into Europe by the armies of Hannibal; but Cuvier soon solved the mystery. He contended that an Indian elephant carried to Siberia could not be changed in the transportation, but would remain an Indian elephant still; and as the bones of the Siberian mammoth showed that it differed essentially in some respects from either spe-

cies of elephant existing at the present time, he averred that it was another species. He averred, farther, that the mammoth was a species belonging to a previous age and now extinct, the proofs of which soon became very abundant from the investigations that were prompted by the teachings of Cuvier. It was shortly after this that the discoveries were made at Montmartre, as noticed in § 383, and these awakened a general interest in palæontological investigations.

408. **Mastodons.**—These animals, one species of which is represented at the right of Fig. 173, differed so decidedly from the mammoths as to constitute another genus. This genus is now extinct, while that of the mammoths, or elephants, still exists, there being now two living species, the Indian and the African elephant, for descriptions of which I refer you to my "Natural History," page 80. The teeth of the mastodon differ very much from those of the mammoth, having the enamel raised up in conical eminences, as seen on the right of Fig. 174, instead of the

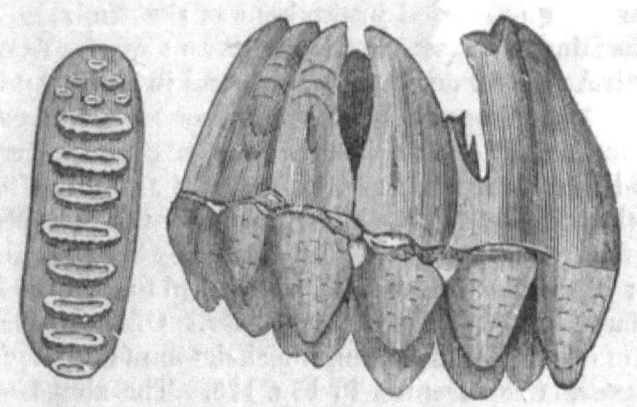

Fig. 174.

arrangement in ridges of the mammoth, as represented at the left of the figure. The *Mastodon giganteus*, an entire skeleton of which is in the Palæontological Gallery of the British Museum, considerably exceeded in size the largest elephant of the present time. Its bones are found only on the North American continent, and are very abundant in some parts, especially in a saline morass in Kentucky called the Big Bone Lick. There is a tradition prevailing among the Indians that there existed men of gigantic stature at the same time with the colossal animals to which these bones belonged, and that both were destroyed by the thunderbolts of the Great Being; but this, like many other traditions, is unfounded, for no bones of man have been found in connection with those of the mastodons. There have been five entire skeletons of this animal dug up in this country. The best one was obtained from a marsh near Newburg, New York, and was set up by Dr. Warren, in Boston. When found, its posture was such as we would expect an animal to have that sunk in mire. Remains of its last meal, lying between its ribs, showed that it lived, in part at least, on spruce and fir-trees. The skeleton is 11 feet high, and its length to the beginning of the tail is 17 feet. The tusks are 12 feet long, over two feet of them being imbedded in the bone of the skull.

Remains of a mastodon of a different species from the North American one have been found in South America.

409. **Mylodon.**—In the pampean or prairie formation of South America, and in the caves of Brazil, have been found the remains of certain huge animals allied to the existing Sloth family. There are only three species of this family at the present time, and they are of moderate size; but of the monsters that belonged to it in the Post-tertiary age there are many species. Of the three species of one genus, Mylodon, the skeleton of one, *Mylodon robustus*, is represented in Fig. 175. The sloths of the present day live in the trees of dense forests, the foliage being their food. The sloths of the olden time lived on the same kind of food, as is shown by their teeth; but they were too heavy to climb, and they obtained the foliage by breaking down or uprooting trees. In doing this, the Mylodon put itself, probably, in the attitude rep-

Fig. 175.

resented in the figure, supporting itself on its two hind legs and its massive tail, as on a tripod. As it had powerful claws, it undoubtedly sometimes dug up the earth around trees preparatory to uprooting them. The animal was of about the size of a rhinoceros or hippopotamus. Its skull has two plates of bone with cells between them, as is the case with the skull of man; but in the mylodon the cellular portion is very large, separating the plates of bone considerably. The reason of this undoubtedly is, that the animal, in tearing down trees, was very liable to have them fall down upon his pate, as his unwieldy form prevented any thing like nimble efforts in getting out of harm's way. If in such an accident the skull were fractured, the break would not be apt to ex-

300                          GEOLOGY.

tend beyond the cellular portion to the inner plate or table, and so the monster's brain would be safe, suffering only a concussion, from which it would speedily recover. The evidences of such a fracture Professor Owen found in one skull, and they were such as to show that the fracture was healed, and that therefore the animal survived the accident.

410. **Megatherium.**—This animal was another of those Post-tertiary South American sloths. Its name, derived from two Greek words, *megas*, great, and *thera*, wild beast, indicates its enormous bulk. The skeleton of one of the dozen species of this genus is represented in Fig. 176. The length of this animal was about twelve feet,

Fig. 176.

and its height was eight. It was exceedingly massive, its thigh bone being twice as thick as that of the elephant, and the other bones are in like proportion. The width of the tail at its upper part was two feet. Its fore foot was more than three feet long, and one foot in width, and the toes were armed with enormous claws. It lived upon leaves and twigs, which it gathered as the

Mylodon did, by pulling down trees, as represented in the case of one species in the front part of Fig. 177.

Fig. 177.

Another genus of the Sloth-like family of the Post-tertiary period was the Megalonyx, the name (*megas*, great, and *onux*, nail) being given to it on account of its large claws. The first species known was found in Virginia, and this name was suggested by President Jefferson. Its size was about that of an ox, but it was much more solidly and heavily built.

411. **Glyptodon.**—This genus, embracing several spe-

cies, all of which were gigantic, was allied to the armadilloes of the present day. One species is represented in the rear part of Fig. 177. This animal had a shell somewhat like a turtle. This solid coat of mail is composed of innumerable plates joined together by serrated or notched sutures. Looked at from within, they are hexagonal, but externally they form a mosaic of rosette-like figures. Remains of this shell were first found in company with bones of the megatherium, and were supposed to belong to that animal. But the truth was soon discovered on finding an entire shell, which is now in the museum of the College of Surgeons in London. This shell is nine feet in length, and the curve, measured across, is seven feet. It is large enough to cover over a large number of the armadilloes of the present day.

412. **Whales.**—These animals appeared in localities in the Post-tertiary period far away from their present habitats. The sea in that age extended up as an arm where the St. Lawrence now runs, and covered the present locality of Lake Champlain. Whales and seals flourished there at that time, and their remains have been found among other spots in the neighborhood of Montreal. In Fig. 178 you have a representation of the bones of the

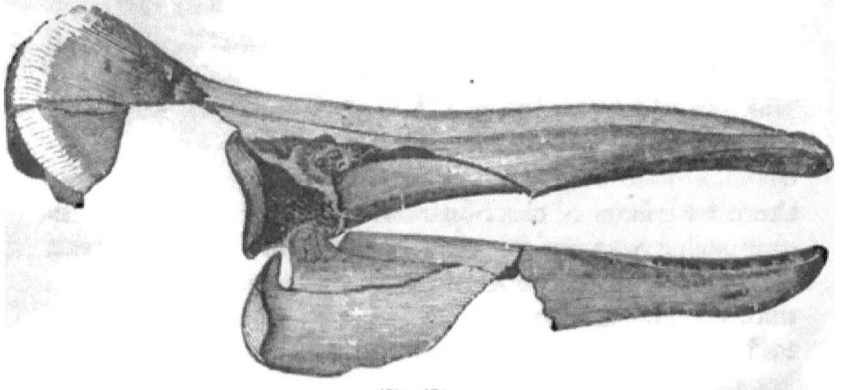

Fig. 178.

head of a small whale, similar to the existing white

whale of the **Northern Sea, as** they were dug up on the borders of Lake Champlain, **60 feet** above its waters, **and 150** above the level of the sea at the present time.

413. **Kirkdale Cave.**—The bones of animals of the Post-tertiary age have been found to some extent in caverns in Great Britain, on the Continent of Europe, and in Brazil, in South America. The European caves were dens of cave-bears, so called because they dragged their prey into these caves. In England they were the dens **of hyenas.** In South America they were occupied by wolves and certain panther-like animals. The most famous cave in England is the Kirkdale Cave, in Yorkshire, the mouth of which was accidentally discovered by some workmen in 1821, in **quarrying stone on** the slope of a limestone hill. I transcribe an account of it given by Agassiz. "Overgrown with grass and bushes, the mouth of this cave in the hill-side had been effectually closed against all intruders, and it was not strange that its existence had never been suspected. The hole was small, but large enough to admit a man on his hands and knees; and the workmen, creeping in through the opening, found that it led into a cavern, broad in some parts, but low throughout. There were only a few spots where a man could stand upright; but it was quite extensive, with branches opening out from it, some of which have not yet been explored. The whole floor was strewn, from one end to the other, with hundreds of bones, like a huge dog-kennel. The workmen wondered a little at their discovery, but, remembering that there had been a murrain among the cattle in this region some years before, they came to the conclusion that these must be the bones of cattle that had died in great numbers at that time, and, having so settled the matter to their own satisfaction, they took little heed to the bones, but threw many of them out on the road with the common limestone. Fortunately, a gentleman living in the neighborhood, whose attention had been attracted

to them, preserved them from destruction; and a few months after the discovery of the cave, Dr. Buckland, the famous English geologist, visited Kirkdale to examine its strange contents, which proved, indeed, stranger than any one had imagined, for many of these remains belonged to animals never before found in England. The bones of hyenas, tigers, elephants, rhinoceroses, and hippopotamuses were found mingled with those of deer, bears, wolves, foxes, and many smaller creatures. The bones were gnawed, and many were broken, evidently not by natural decay, but seemed to have been snapped violently apart. After the most complete investigation of the circumstances, Dr. Buckland convinced himself, and proved to the satisfaction of all scientific men, that the cave had been a den of hyenas at a time when they, as well as tigers, elephants, rhinoceroses, etc., existed in England in as great numbers as they now do in the wildest parts of tropical Asia or Africa. It was evident that the hyenas were the lords of this ancient cavern, and the other animals their unwilling guests, for the remains of the latter were those which had been most gnawed, broken, and mangled; and the head of an enormous hyena, with gigantic fangs, found complete, bore ample evidence to their great size and power. Some of the animals, such as the elephants, rhinoceroses, etc., could not have been brought into the cave without being first killed and torn to pieces, for it is not large enough to admit them. But their gnawed and broken bones attest, nevertheless, that they were devoured like the rest; and, probably, the hyenas then had the same propensity which characterizes those of our own time, to tear in pieces the body of any dead animal, and carry it to their den, to feed upon it apart."

## CHAPTER XIX.

### AGE OF MAN.

**414. Boundaries of this Age.**—In the ages of the far past, as you have already seen, there is no exact line dividing one age from another. So it is with the present age, and the Post-tertiary that preceded it. It has been the common idea that the advent of man is fixed by the chronology of the Bible at about 6000 years ago, and the researches of geology have been thought to coincide with this; but recently there have been some researches which seem to put the introduction of man farther back than this. The question is yet undecided, but if the result indicated should be arrived at definitely, it would not show that the Bible is false, as some would have it, but merely that the common view of its early chronology is wrong. As to the conclusion of the present age, geology leaves us entirely in the dark. The Bible does indeed point to a time when the affairs of this world shall be concluded, and the earth cease to be the habitation of the human race, a vast change being indicated by the announcement that " the heavens shall pass away with a great noise, and the elements shall melt with a fervent heat; the earth also, and the works that are therein, shall be burned up." But the geologist, though he sees in the operations that are now going on evidences of great future changes in the arrangement of the earth's surface, finds nothing which could enable him to predict any such radical change as the Bible plainly, though in general and indefinite terms, points out.

**415. Earth now and in Former Ages.**—The contrast between the state of the earth in the present age and in any of the former ages is very great, and the farther we go back in the comparison the more striking does it appear. To say nothing of that period or age when the earth was in a fused condition, look at the Azoic age as compared with the present. When the extent of land was small, and there was no life either on land, or in water, or air, there were no high mountains, and no rivers of any size. It was a dull, monotonous world, compared with the variegated earth which we have now, with its continents, islands, mountains, lakes, and rivers, all swarming with busy and noisy life. A strong contrast can be made out in regard to the ages that followed, especially in relation to the changes of various kinds that were necessary in making the requisite additions to the continents, such as flexures, upheavals, denudations, deposits, elevations, subsidences, etc. Disturbances, it is true, are occurring in this age, many of them precisely of the same character; but they are not so great, nor so wide in their range. For example, glaciers and icebergs are at work now; but their work is not continent-wide, as it was in the Glacial age, when ice reigned supreme over a large portion of the earth, in order to prepare it, as you have seen, for man. So, in the formation of peat, we have the same thing essentially as was the grand business of the Carboniferous age; but it is a small operation compared with the coal-making of that period. The Post-tertiary period was the nearest in character to the present, especially the Terrace epoch, for the continents were then finished, and all the grand general work of diversification of the surface was completed, only the minor diversification, as it may be termed, remaining to be done.

416. **Completion of the Earth.**—Although changes, and those of no small extent, are now going on in the crust of the earth, yet, in a certain sense, it may be regarded as having been completed at the time when man was introduced upon it. This conclusion is seen to be true if we consider certain prominent facts, viz., that the conti-

nents had their germs, and grew gradually by accretions, after a regular plan, to their present dimensions; that after their proper size was reached great operations were instituted to diversify the surface, and to grind up some of the rocks into soil for the use of man; that the preparations of the earth's crust, even down to minute circumstances, so far as they are understood, can be seen to look toward the consummation arrived at in the age of Man; and, finally, that man, appearing after a succession of animals extending through long ages, differs from them all in the possession of qualities that ally him with the Infinite, and therefore show him to be the fitting end of such a consummation. The earth was made for man. Accordingly, he has a general control over the powers of earth, being commissioned " to subdue it, and have dominion over the fish of the sea, and over the fowl of the air, and over every living thing that moveth upon the earth." Mind gives him this control, even where physical circumstances, as weight and strength of muscle, are sufficient to overpower him in any direct effort. He is the lord of this lower creation, acting as the vicegerent of the Creator.

417. **Changes now Transpiring.**—The changes which have taken place during the age of Man, and are now going on, have been quite fully noticed in Chapter X., and I shall only give a summary of them here. The rocks are every where subjected to weathering, which wears them away by piecemeal, and the rivers are carrying away the detritus of the rocks, lodging it in deltas or on flood-plains. The tendency of all this is to fill up the depressions, and to make the land encroach upon the space occupied by the water. Then there is evidence that over extensive areas of the floor of the sea deposits are being laid down of the same materials that now compose rocks made in former ages. Besides, there are the coral animals busily engaged in building extensive reefs. There is, therefore, great land-building going on, and the

result must inevitably be, that the level of the ocean will rise, unless there be some movements of the land to counteract it. Elevations in the land or depressions in the floor of the sea would do this. That elevations have occurred within this age, and are occurring now, we have the clearest evidence, as shown in Chapter X. It may be that the heaving of volcanoes, adding to the land, may occasion some corresponding depressions in the floor of the sea in their neighborhood. At any rate, in some way the general level of the ocean in relation to the land is maintained about the same from century to century. The preservation of this equilibrium in the midst of constant changes is secured by the same all-wise and all-powerful Providence that preserves the universal constitution of the atmosphere, as noticed in § 85, Part II. This appears very wonderful when we reflect that the earth is essentially a molten ball covered with a hard crust, which has vent-holes communicating with the internal fires, and contains vast quantities of water in its depressions.

**418. Present Activity of the Agencies Producing Change.**—The causes of the present changes in the earth were brought to view in § 172, and were there stated to be the same as have been in action in the different ages of the earth's formation. All of them have been in operation from the first except one, that is, life, which was introduced upon the stage after the long periods of the Azoic age were passed. An interesting question arises in regard to the degree of activity of these agencies at the present time. Lyell and some others claim that they are as active now as they ever were; but the evidences in favor of the opposite opinion are very decisive. They surely are not as strongly at work now as they were when they raised up the lofty mountains, or when, in the Azoic age, the crust of the earth was solidifying at the first, and the germs of the continents were forming.

**419. Historic and Human Periods.**—What is called the

Human period covers the whole age of Man, but the Historic period does not. What purports to be the early history of many of the old nations is made up of traditions, which are far from being reliable. Real history, with the exception of the inspired history of the Bible, goes but a little way back in the case of any ancient nation. "Milton did not scruple to declare," says Hume, "that the skirmishes of kites or crows as much merited a particular narrative as the confused transactions and battles of the Saxon Heptarchy." But the study of the remains of a people may be carried far back into the past, beyond the beginning of that degree of civilization which is necessary to authentic history, and may give us valuable results. It is by such study alone that we can acquire a knowledge of ancient nations in their savage or barbaric state.

420. **Stone, Bronze, and Iron Ages of Man.**—In marking the progress of man by means of the remains of his implements, weapons, ornaments, etc., the materials which he has used in making them have been considered as denoting three stages. In the first and rudest condition we have the *Stone* age, when stone was the material from which were shapen such implements as hammers, axes, chisels, and arrow-heads. In Fig. 179 (p. 310) are represented some of these early stone implements gathered from various quarters, all of them exhibiting a striking similarity, but some of them being a little more elaborate than the others. Those at 1 and 2 are from the Valley of the Somme, in France; at 3, 4, and 5, from England; at 6, 7, and 8, from Canada; and at 9 and 10, from Scandinavia. Quite an advance upon this is the *Bronze* age, for there are more thought and skill shown in melting metals together for the making of implements than in the bare shaping of stone. Copper was the chief metal of this age. Last of all comes the *Iron* age, in which the excellence of iron above all other metals for the manufacture of tools and weapons is

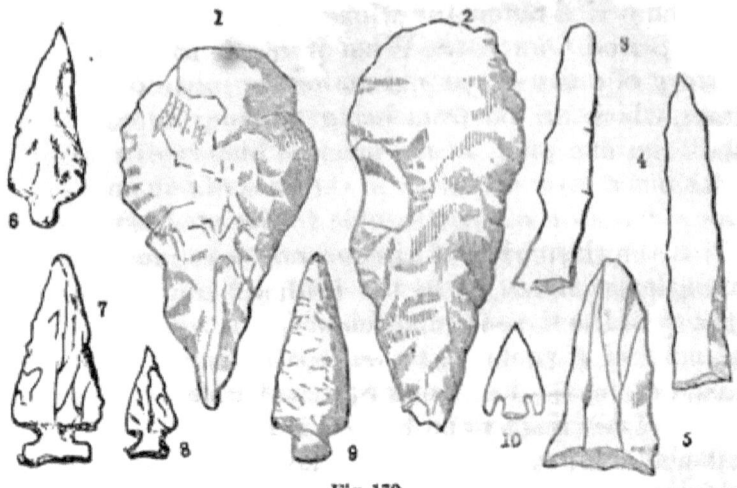

Fig. 179.

recognized, the advance in thought and contrivance having enabled man to discover the art of preparing it, by various smelting processes, for this purpose. It is curious to notice how each of these ages has its peculiar style of ornaments, dwellings, etc. There is no literature, and therefore no history, in so rude a condition as that of the Stone age, and almost none in the Bronze; but when a nation or tribe is so far advanced in the arts as to work iron, there is a literature, small at first, but increasing with every advance in the arts of civilization.

421. **Animals.**—In the age just previous to the age of Man brute force held sway in the animal kingdom. It was fitting, therefore, that as the reign of intellect in the world was ushered in, such monsters as the megatheriums, mammoths, mastodons, etc., should drop out of existence, and that the animals so useful to man, and so easily controlled by him, as the ox, the horse, the sheep, etc., should appear upon the earth in such abundance. The line of separation between this and the Terrace period in regard to animals is not a fixed and definite one. It is not so even in regard to the Mammalia. Some existing now began their existence in the Terrace period,

but there is a difference of opinion as to how many of them did so. There are more of insects and birds in the present age than in any previous one, but not so many mammals as in the Mammalian age, or reptiles as in the Reptilian age. Some animals have become extinct within the memory of man. Outlines of three of them are given in Fig. 180 (p. 312). The shortest one, the *Dodo*, which lived in Mauritius and other adjoining islands, was a heavy, clumsy bird, covered with loose, downy feathers, and having imperfect wings. It weighed about fifty pounds. The earlier voyagers saw it, and made sketches of it; but after the possession of the island of Mauritius by the French in 1712 it was no longer known, and one or two heads and feet of this bird are all that remain of it in the cabinets of Europe. The tallest figure is an outline of the *Dinornis elephantoides* of New Zealand, exceeding the ostrich in size. The name comes from two Greek words, *deinos*, terrible, and *ornis*, bird. The outline on the right represents another species, the *Dinornis ingens*. Bones have been found in Madagascar similar to those of these birds, and quite as large, and with them some egg-shells. The bird to which they belonged has been called *Æpiornis maximus*. Its egg was over a foot in diameter, and equaled 148 hen's eggs and 50,000 humming-bird's eggs in size.

422. **Man One Species.**—Although some few physiologists hold an opposite opinion, the evidence is very decided that all the varieties of the human race belong to one species, as is declared in the Bible. As this evidence is brought out quite fully in my "Human Physiology," I will not go into details here. Suffice it to say now, that the resemblance, we may say identity, of the races of men in all essential physical characteristics, and especially in those which are mental, is unmistakable, and that all the varieties can be accounted for from the influence of circumstances, which, indeed, acting also upon other animals inferior to man, but every where ac-

Fig. 180.

companying him, as dogs, horses, etc., have produced in them somewhat similar results.

**423. Man's Place in Nature.**—The grand difference between man and other animals is one of *kind*, and not of degree only. He is not merely the highest of a series of animals, as some assert, but he possesses certain qualities that do not belong in any degree to the animals below him. He stands in some respects alone. In physical structure he is, indeed, allied to other animals, because he lives amid the same material circumstances, and has similar bodily wants. These relations with other things make it also necessary that he should have similar instincts, and, to some extent, similar thoughts and reasonings, for brutes do have a lower order of reasoning—that is, they draw simple inferences. But here the resemblance stops. There is a certain department of mind which belongs exclusively to man, and separates him by an impassable gulf from other animals. It is the power of abstract or general reasoning that distinguishes the mind of man from that of the brute. No brute, however much he may know, *can ever himself either discover, or receive by instruction, any general principle*. It is this power that gives man the knowledge of the existence of a great First Cause, and of the difference between right and wrong, and that introduces him into a sphere of thought and feeling which he occupies in common with the angels, and with the Creator himself. It is this which makes him "a living *soul*." It is from this that he is said to be created in the image of God. From this come all his endless contrivance in implements and machinery, his attainments in science, and his construction of language. This subject, thus briefly noticed here, is fully treated in my "Human Physiology."

424. **A Supposition.**—It is the idea of some that, as man is the present culmination of the animal kingdom, there is still to be an onward progress, and that some other being of a higher order even than man will, after a while, appear upon the scene. There would be some reason for this expectation if there had been a regular gradation from the first dawn of life in the Silurian age,

O

evolving man at last at the summit of the series, the difference between him and the highest of other animals being only in degree. But there is no such regularity of gradation, and man is not merely the highest in grade of all animals, but he differs from them in the possession of mental attributes, that ally him with the Infinite, and force upon us the conviction that the earth was made for him, and that he alone is the end of its creation.

## CHAPTER XX.
### CONCLUDING OBSERVATIONS.

**425. Geology and Astronomy Compared.**—In Astronomy we take into consideration systems of planets which are at immense distances from each other, but in Geology our view is confined to a single planet, and that a comparatively small one. In the study of Astronomy we have only the vast and the grand before us, and there is nothing small, or even moderate in size, much less minute; but in Geology we have a combination of the vast and the small, so that the study has more of the elements of interest in it than Astronomy presents. While conceptions of grandeur are sufficiently awakened by some of the extended movements of continent-making, much of the subject lies directly under our feet, and invites the most familiar examination. The interest is enhanced when we see that even operations so minute as to call for the microscope in their investigation have been concerned in laying down strata of immense extent and thickness, in the building up of continents, and that these operations are in some degree still going on. Besides, the fact that it is *our* earth that we examine in its vast stony leaves—that all those multiform processes which have occupied long ages were constructing a habitation for us, gives to the study an interest which does not attach to the study of other planets.

**426. Record of the Rocks and History.** — In reading history as written by men there is always more or less doubt about its reliability. It is even so with the history of recent times, but more especially with that of times long gone by, as already alluded to in § 419; but the record which the Creator has left in the rocks is a true history. When Geology first unfolded its leaves to the world, it was the idea of some that the animal and vegetable forms found in the strata were the mere images of living things imprinted there, and not actual remains. This most unworthy idea of the Deity's work of creation is not now entertained by any one, but the history folded up in the rocks is universally recognized as a true life-record of past ages. We may sometimes err in reading it, or may even fail to decipher it; but, nevertheless, there is no mistake in this record written by the Infinite and the True.

**427. The Two Divine Records of the Creation.** — There are two authentic records of the creation of the earth, the one contained in the Bible, and the other inscribed on the rocks. Some scientific men, who do not absolutely deny the truth of the Bible, seem to think that the record developed by their discoveries has a certain and indisputable claim on their faith, which the other record has not. But both are equally authentic, though the evidences of authenticity are different in the two cases. It is just as well established that the Bible is a divine record, the authors being merely agents of the Deity, as that the record in the rocks was made by divine power, heat, water, light, electricity, etc., being the agents by which it was made. These two records can not, then, be inconsistent with each other, though they may be apparently so, from a wrong interpretation of the one or the other. In interpreting the account of the earth's creation in the Bible, we must remember that it does not purport to be a scientific account, and therefore common, and not scientific expressions are used. Judged of in a reasonable manner, it is plain that this account, which is so universally admired for its sublimity, has remarkable coincidences with the record in the rocks. These have been admirably traced out by Hugh Miller, Dana, and others, and I will not dwell upon them here. They are such that every candid mind must conclude, in view of them, with Dana, that "no human mind, in the early age of the world, unless gifted with superhuman intelligence, could have contrived such a scheme." It might have been done by some very wise man of the present time, with the aid of all the knowledge which the researches of Geology could give him, though not with such remarkable skill and sublimity; but to do it previous to the acquisition of this knowledge would be an impossibility. The conclusion, then, is inevitable, that Moses was guided by superhuman power in making the account, or, in other words, was divinely inspired.

Many geologists think it out of place to notice at all the record in the Bible; but if it be true that the Deity has given us a written record, it is our duty to examine it thoroughly and candidly, and a reasonable reference to it in a scientific work can not be out of place. It is the dictate of science as well as religion to notice it, and therefore it is done by such eminent men as Dana, Hitchcock, Hugh Miller, etc.

428. **The Days in the Mosaic Record.**—That the days which Moses says were occupied by the Creator in the creation were long ages is evident from a comparison of the order of events in that narrative with that revealed by the record of the rocks. But perhaps it is objected that the plain reading of the account would lead any one to believe that the six days of the creation were days of twenty-four hours each. So it would, if not interpreted by the knowledge which Geology has given us, just as we should infer from the language of the Bible that the sun actually rises and sets, as the ecclesiastical court that imprisoned Galileo believed, until Astronomy taught us

otherwise. In regard to the word day, we continually use it in various senses, and Dana states that it is used in five different senses in the Mosaic account of the creation. "'These are,' he says, (1.) The light—'God called the light day,' v. 5; (2.) the 'evening and the morning' *before* the appearance of the sun; (3.) the 'evening and the morning' *after* the appearance of the sun; (4.) the hours of light in the twenty-four hours (as well as the whole twenty-four hours), in verse 14; and, (5.) in the following chapter, at the commencement of another record of creation, the whole period of creation is called 'a day.' The proper meaning of 'evening and morning' in a history of creation is *beginning* and *completion;* and in this sense darkness before light is but a common metaphor." It is a significant fact, I add, that the word day is applied to the three first periods of creation, when as yet the sun had not appeared. There could have been then no such division as our day; and, indeed, it is expressly stated that this division was introduced in the fourth period or day of creation, for it is said of the lights in the firmament, "Let them be for signs, and for seasons, and for *days*, and years."

429. **Traditions and Superstitions.**—There have been many traditions and superstitions in regard to creation, and the origin of various fossils that were accidentally found in the rocks before geology was known as a science. I have here and there referred to some of these. There is an interesting English legend in regard to fossil ammonites. These abound in the neighborhood of Whitby, in Yorkshire, and it was a common belief there that they are petrified snakes. The story of their petrifaction is this: As the snakes were so numerous as to prove a great annoyance to the inhabitants, they implored their patron saint, St. Hilda, to intercede for their destruction, whereupon she not only prayed their heads off, but prayed them also into stone. Sir Walter Scott thus records the legend in his Marmion:

"And how the nuns of Whitby told
How, of countless snakes, each one
Was changed into a coil of stone
When holy Hilda prayed.
Themselves within their sacred bound,
Their stony folds had often found."

So lately was this superstition current, that the author of a modern scientific work relates that a sharp dealer, who was requested by his customers to supply them with some of the creatures that had escaped that part of St. Hilda's prayer which destroyed their heads, affixed to the fossils some heads of plaster of Paris suitably colored. He had a thriving trade till it was upset by some officious geologist. But even now the fossils are sold in Whitby with the extremity of the last whorl filed into the shape of a snake's head, as represented in Fig. 181.

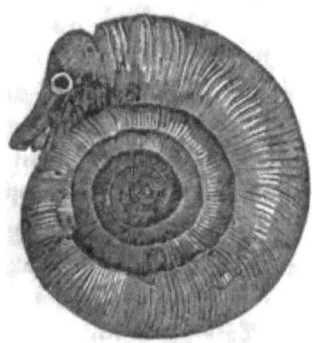

Fig. 181.

**430. Use of Geology to the Poet and the Painter.**—Geology is of service to the poet in adding largely to his fund of facts such as are eminently fitted to awaken the sublimest thoughts and feelings, and to give a wide range to the flights of his imagination. It is of use to the painter of scenery as anatomy is to the painter of the human form, for it gives us, as we may say, the anatomy of the earth. Besides, it supplies him with many valuable and interesting hints as to the objects upon the earth's surface.

**431. Plan in Earth-development.**—You have seen, in the course of your study in this book, that the Creator worked from the beginning after a plan, in developing the continents on the crust of the earth, and if we could go down into the depths of the ocean, and examine the irregularities of its floor, we should undoubtedly see the same thing there. What the geologist has shown us in regard to this plan is like what we see as we look upon a potter as he makes some vessel. We see the general plan developed very soon after he begins to turn the ball of clay, and the minutiæ of the plan appear more and more as he proceeds. Just so the investigations of geologists have shown how the continents, at the very commencement of their formation, had a shape which looked toward what they now are, and how they were gradually developed, and at length were finished in all the minutiæ of their diversification. And I may remark here that each continent had its own plan. While, for example, the North American continent, as you saw in § 267, began with the formation of one long island, Europe was at first a group of islands, which were afterward united together. There was a plan, also, in regard to the animals of the earth. This is seen in the preservation of the four grand divisions of animals from the beginning, through all the changes of genera and species; in the gradual advance from the lower up to the higher forms, through the ages of the earth's growth; in the introduction of particular forms for special purposes, at certain periods, and in certain localities; and in the final consummation of the animal kingdom, in which, after all its long line of successions, there was an adaptation of it to the wants of man, who was constituted its ruler. The same may be substantially said of the vegetable kingdom.

432. **Time in Geological Processes.** — I have here and there, in previous chapters, given some illustrations of the great lengths of time required for the formation which make up the crust of the earth. You will recollect the calculation of Liebig, noticed in § 318, in regard to the formation of coal. Calculations have also been made in regard to the deposition of the strata of rocks, and the estimates for all the formations, from the Azoic down, reach a sum total of over fifty million of years. The estimates are based upon the rate at which rocks are deposited and solidified in lakes and seas at the present

time. There is nothing in science which so well impresses the mind with the truth of the declaration of the Bible, that "one day is with the Lord as a thousand years, and a thousand years as one day."

433. **Minute Agencies.**—I have brought to your notice in this book many examples of extensive operations in the construction of the earth by minute agencies. It is a very small thing for a coral animal to separate from the water a little carbonate of lime and appropriate it to itself; but multitudes of these animals, at work in the same locality, year after year, for centuries and even ages, lay down thick masses of limestone rock. The case is even stronger with the diatoms, microscopic vegetable organisms which separate silex from its solution in the water and deposit it in thick beds. A large portion of the crust of the earth is, in fact, the result of the aggregate labor of minute animals and vegetables.

434. **Disintegration of the Rocks.**—In the preparation of the earth for the use of man, the disintegration of the rocks has been a prominent process. The soil which he cultivates is, as you have seen, a result of this disintegration. The mud that is carried down the rivers, and deposited in deltas or on flood-plains, is comminuted rock gathered by the waters from mountains and hills. The ice, in glaciers and icebergs, is continually grinding up the rocks to add to the soil of the earth. A vast work has been done in the past by water, in both its liquid and solid forms, in this preparation of the earth for vegetation. This was done largely, as you have seen, in the Glacial period; but the work has always been going on, for water has always been in motion.

435. **Reconstruction.**—In the building up of the earth there has been a vast amount of reconstruction. The rocks of which the crust of the earth is composed are made, to a very great extent, of materials derived from the ruins of rocks previously made; and often the materials have been used over and over again. Disintegra-

tion, then, has not been of use solely or chiefly in preparing the soil for vegetation. Its principal use has been to procure the proper arrangement and condition of the earth's crust. You have seen how largely, in some cases, denudation has been carried on. This is chiefly for the purpose of removing the material to localities where it is required. It is by this reconstruction, so extensively prosecuted, that the proper diversification of the surface of the earth has been effected.

436. **Mechanical, Chemical, and Vital Agencies.**—In all these disintegrations and reconstructions you see a great interplay of mechanical, chemical, and vital forces. I will cite here a single example. Carbonate of lime, in the forms of limestone, chalk, and marble, enters largely into the structure of the crust of the earth; but water is continually dissolving some of this as it finds it in the rocks, and so the ocean is kept supplied with it. This is a *chemical* operation, the carbonic acid in the water enabling it to dissolve considerable of the carbonate of lime. Then, following it to the ocean, shell-animals there gather it to make their shells, or coral animals to make their skeletons. This is a *vital* operation, for in the bodies of these animals, by a vital power, the carbonate of lime is separated from the water in which it is dissolved, and is deposited in a solid form. Then, by a *mechanical* operation, these shells and skeletons become massed into solid rock.

437. **The Atmosphere.**—One of the most signal examples of the interplay above referred to is seen in the relations which the atmosphere bears to the earth that it envelops like a robe. Before the Carboniferous age it was a very different atmosphere from what it is now. It was then highly charged with carbonic acid gas, which is carbon, or charcoal, united with oxygen. Now this excess of carbon in the air was brought down in the Carboniferous age, and lodged in the bowels of the earth, as coal, in immense stores, for the future use of man.

Observe how the mechanical, chemical, and vital forces were in operation here. The carbon of the air was made a part of the wood in the enormous vegetation of the Carboniferous era, through the vital and chemical action of the leaves (§ 81, Part II.). Then, by chemical and mechanical operations together, this wood was accumulated under strata and changed into coal. The same operations are going on now, but not to any thing like the extent that they did in the Coal-making age. For ages, then, we may say, the coal which mankind are now burning was stowed away in the atmosphere, unseen in its gaseous union with oxygen, and at the proper time it was brought down, and stowed in solid form in the bowels of the earth, ready for man's use.

**438. Circulation of Matter.**—You see that by the forces which I have mentioned a large portion of the matter in the world is in constant motion, water being the chief, but not the only agent by which the motion is kept up. More of it was in circulation in former ages, when the continents were in process of formation, than there is now, when they are, in a certain sense, completed. There is enough of it now remaining quiet to give "the foundations of the earth" stability, and the great object of the present circulation of matter is to keep up the operations necessary to make the earth a fit habitation for man.

**439. Exertion of Creative Power.**—It is the notion of some that the Creator put a certain amount of matter into this ball that we call earth, and then left it, under the operation of certain laws or tendencies, to work out all the results which we see. In other words, the machinery, mechanical, chemical, and vital, was wound up, so to speak, in the beginning, and has not been touched since by the creative hand. That there have been no additions to the matter in this world since it was first launched into space is probably true; but that creative power has often been exercised in giving new properties and tendencies to matter there is the most decisive evidence. Geology reveals the fact that there was once a long age in which there was no life, vegetable or animal, upon the earth. After this was passed life was created —that is, by creative power there were given to portions of matter properties which matter of itself can never originate. In fact, every new species was an independent creation. To assert, as some have done, that there never was any creative power exerted, and that all we see is the result of the fortuitous concourse of atoms, is a folly which does not merit a moment's notice.

440. **Development Theory.**—There is a class of unfounded theories that have been broached from time to time in regard to the origin of species, all of them bearing essentially the same character. Among these, one recently brought out by Darwin, called the development theory, is just now attracting much notice. He supposes that, at the outset, there was "a breathing of life into a few forms, or one form," and that all the living forms which we see were evolved from this beginning by what he calls "natural selection in the struggle for existence." He admits, therefore, a divine creation of life at the first, but claims that the development of all living forms after this was directed by *chance* and *Nature*. All species, in his view, came from mere varieties that were started in this "struggle." In evolving thus one species from another, it is supposed that there were *intermediate* forms, and that much time was required for the passage through the gradations. Unfortunately for this part of the theory, geology steps in with a most complete refutation from its life-record in the rocks. *Not a single intermediate form has been found in all this record.* If they had ever existed they must have been found, and that abundantly in the midst of the multitude of species that appear. The changes demanded by this theory in converting one species into another are great changes. The matter is well stated by Page thus: "Given the scales, fins, and gills of a fish—what the conditions, and what

the amount of time necessary to transmute them into the scutes, paddles, and lungs of a marine reptile? Given the scutes, membranous forearms, and stomach of a flying reptile—what the phases of change, and what the amount of time required for their transformation into the feathers, wings, and gizzard of a bird? Or, given the four hands, with partially opposable thumbs, the low facial angle, and the jabbering, half-reasoning of a monkey—what the force of conditions, and what the term of time for their development into the two-handed dexterity, the erect aspect, and the eloquent ratiocinations of a philosopher of the nineteenth century?"

There are many fundamental objections to this theory, but this is not the place for a full notice of them.

441. **Creation of Man.**—Such writers as Darwin and Huxley seem to think of creation as only a creation of *material* forms, and do not indulge in the least the idea of a spiritual creation. But, in truth, *the grand event in the successive creations of which this earth has been the scene is the creation of a living* soul. Man is not a mere congeries of living organs, endued with certain properties, but a spiritual existence connected with such a congeries. Now, that this soul was created as really as the congeries of organs with which it is united appears from two facts: 1. It has certain attributes which, as stated in § 423, show that it is to some extent different, not merely in degree, but in kind, from the mind of brutes, and therefore had an independent origin. Not only is there no proof that the mind of man was generated from the mind of a monkey, as claimed by Huxley, but there is positive proof to the contrary in this radical difference. 2. This proof is enhanced by the fact that there is no difference in brain that corresponds at all with this mental difference. This utterly refutes the materialistic doctrine that mind is a mere product of matter, and that the differences of mind come from differences in material organization; for, if this were so, we should rightly ex-

pect (as I have stated in my "Human Physiology," § 523) that, as the mind of man is specifically different from that of the highest of the inferior animals, there would be a specific difference in the brain. The fact that there is no such difference shows that the mind of man is, in a certain sense, independent of material organization, and was a separate creation, though their connection establishes an intimate mutual dependence in man's present existence in this world. What folly, then, is the doctrine of a "science, falsely so called," which places man only a little above the monkeys, when the Creator has declared in his Word that he "made him a little lower than the angels," and a true science agrees with the declaration!

# QUESTIONS.

[The numbers refer to the pages.]

## CHAPTER I.

9. What is said of the forms of minerals? In what sense are all minerals solid? Illustrate the relation which heat has to the forms of matter.

10. What are the three kinds of substances that are not mineral? State what is said about the lifeless period of the earth.

11. What is said about the introduction of life? What of the different results of chemical and of vital action? Explain what is called decay. Give in full the illustration in regard to the egg.

12. What is said of mineral matter in regard to organic substances? Of what materials have many rocks been made? What is said of carbonate of lime? What of the flinty rocks?

13. Give examples of simple minerals. What metals are never found uncombined with other elements? What metals are found both native and in combination? With what are they combined? What is said of oxygen? Of hydrogen? Of chlorine? Of sulphur and carbon? What is said of the different degrees in which minerals are compound? Mention the composition of mica and of lapis lazuli.

14. What is said of granite as a mixture of mineral compounds? State the relations of Mineralogy, Chemistry, and Geology.

## CHAPTER II.

14. What takes place in crystallization? What are familiar examples of crystals? What is the arrangement of the crystals of mica?

15. What is said of perfect and imperfect crystallization? What is an octahedron? What is said of the sizes of crystals?

16. State fully the difference between the mode of crystallization and of vital growth. What is said of crystals with curved lines? What of spherical forms in some rocks? What of the arrangements of crystals?

17. What is said of the manner in which mineral matter is deposited in living substances? What vegetable substances assume crys-

talline forms? Under what circumstances do they do this? What is said of forming crystals by deposit from solution?

18. Explain the formation of frost and dew. Give examples of the conversion of liquids and gases into crystals. What is said of the production of crystals by the influence of heat? What of the water of crystallization? What is an amorphous mineral? What a dimorphous mineral?

19. What is said of the arrangement of crystals? What of the crystals of common salt? What arrangements are there in rocks that are somewhat like crystallization? What is cleavage? What are some of its different modes?

20. What is said of cleavage in quartz? What are primary forms? How many are there? Into how many classes are they divided? What are the crystals of the first class? What is a rhomb? What an octahedron? What a dodecahedron?

21. Show how the cube may be converted into the octahedron by cleavage. Show how the octahedron can be converted into the cube. What are secondary forms? In what two ways are they produced in nature? Illustrate the mode by addition.

22. What is said of constancy in the forms of crystals? What of symmetry?

22. Illustrate the analogy between the symmetry of the mineral and that of the living world.

## CHAPTER III.

23. What are the four grand elements of which animal and vegetable substances are composed? What is said of the diffusion of carbon? What constitutes the diamond?

24. How can the diamond be proved to be pure carbon? How does it differ from charcoal and anthracite? What is said of the coloring of diamonds? What are the qualities of diamonds? What is said of their size? What are the meaning and origin of the word *carat?*

25. What is said of the cost of diamonds? What of the art of cutting them? What of the expense of the process? What of the uses of the diamond?

26. What are the chief localities in which diamonds are found? What is said of the modes of collecting them? What is the composition of graphite? What are its properties? What its uses? How is it prepared for use in pencils? What is the difference in composition between bituminous and non-bituminous coal? What gas is produced from bituminous coal?

27. What is the blue flame of imperfectly burning anthracite? How

is illuminating gas obtained from bituminous coal? What is coke? How is it supposed that anthracite coal was produced? How is it like coke? How does it differ from it? What is its ordinary composition? From what does redness in the ash come? What is said of slag? Why do oyster-shells thrown into a coal fire remove the slag or clinker? What is said of certain uses to which anthracite is sometimes applied?

28. What is said of the varieties of bituminous coal? What of the mineral called jet? What is lignite? What proofs are there that coal is of vegetable origin?

29. How does carbon as it is in wood differ from carbon as it is in coal? Give the comparison in relation to its combination in wood. Give in full what is stated in regard to the change in producing coal. What is peat, and how is it made?

30. What is said of the depth and extent of peat beds? What of the diffusion of coal in the earth? What is amber? What are its properties? What its uses? What is said of specimens of it and imitations of them?

31. How are the different kinds of bitumen produced? What is said of asphaltum? What of petroleum?

32. What of naphtha? What is the composition of carbonic acid gas? What are its qualities? From what sources is it supplied to the atmosphere? How is its undue accumulation in the atmosphere prevented? What is said of localities where it is produced in large amount? State in full what is said of its agency in relation to limestone. What is said of the natural production of illuminating gas in some localities?

## CHAPTER IV.

33. What is said of the crystallization of sulphur? What of its localities? From what ores is much of our sulphur obtained? What are some of its uses?

34. What is the common iron pyrites, and why is it so called? What are the forms of its crystals? What mistake is often made about it? How is it distinguished from gold? What use is made of it? What are the two other kinds of iron pyrites mentioned? What is said of copper pyrites? What is copper glance? What is said of a sulphuret of a very compound character?

35. What is galena? What is said of it? In what parts of the United States is it found abundantly? What is said of the sulphuret of silver? What of the sulphuret of antimony? What of the sulphurets of arsenic?

36. What of the sulphuret of mercury? What of the sulphuret of

zinc? Give the names and composition of the three vitriols. How does a sulphate differ from a sulphuret? How are these three sulphates produced from the sulphurets of the same metals? What is copperas? What are its uses? What is said of sulphate of lead?

37. What is the composition of gypsum? What are its properties? What is said of some of its forms? What is anhydrous gypsum? What is said of the gypsum found in the Mammoth Cave of Kentucky? State the case related by Prof. Hitchcock, and the mode of detecting the error indicated.

38. What is said of the sulphates of magnesia and soda? What of the sulphate of baryta? What of sulphuric and sulphurous acids? What of sulphureted hydrogen?

## CHAPTER V.

39. What are *native* metals? To what is the term ores usually applied? In what different senses is it used? What is said of the positions and associations of ores?

40. Illustrate the simplest method of obtaining metals from ores. Illustrate the other method mentioned. What is the *gangue* of an ore? What is the process termed *washing?* Illustrate the operation of *fluxes*. What are the most common ores of iron? What is said of them as colorers of rocks and soil?

41. What is said of meteoric iron? What of magnetic iron ore? What of hematite? What of brown iron ore?

42. What of chromic iron? What of carbonate of iron? Under what circumstances is native copper found? What is said of finding silver with it? Where are famous copper mines? What is said of the copper region of Lake Superior?

43. How was the copper produced in that region? What is said of the oxyds of copper? How many carbonates of copper are there? What is said of the one called malachite? What of the silicate of copper? What of native lead? What is red lead? What white? What chrome yellow? What is said of tin and its ores?

44. What is said of zinc and its ores? What of antimony and its compounds? What of the ores of cobalt? What of nickel and its ores? What of bismuth and its compounds?

45. What is said of the ores of manganese? What are the properties of mercury? What is related of its discovery? Why is it called quicksilver? How is its purity estimated by the miners? What are its uses?

46. In what various conditions and combinations is silver found? In what states is gold found? What is the only compound that has yet been met with? How is gold distinguished from iron and copper

pyrites? In what different forms is native gold found? What is said of some of its localities?

47. What is said of the amount of gold obtained from Australia and the United States? What of the modes of obtaining gold? What of the sorting of gold by natural operations? Explain the operation by which a "pocket" of gold is formed.

48. Describe the manner of effecting a large "washing" in the search for gold. What are the quantities of gold that make it so useful? What is said of gilding? What of the alloys of gold? With what is platinum generally found associated? What are its properties?

49. What is said of the metals found in company with platinum?

## CHAPTER VI.

49. What are oxy-salts? What are haloid salts, and why are they so called?

50. What is saltpetre? What is said of its production in nature in different localities? Of what use is it in the composition of gunpowder? What is said of the nitrate of soda? What is the composition of common salt? What is said of its diffusion in the earth? Under what circumstances is water in seas or lakes exceedingly salt? Where are there famous salt mines? What is said of the mines of Cracow? Where are there hills of salt? How is salt obtained in this country?

51. What is the composition of borax? What is said of the most noted locality of it? Under what circumstances is boracic acid evolved? What are the different forms of carbonate of lime? What is said of its diffusion in the earth? How is it distinguished from other minerals that resemble it? What effect does heat produce upon it? What is hydraulic lime?

52. What is marl? What calcareous tufa? What fact is stated about chalk? To what is the name calcareous spar applied? What is said of some of its varieties?

53. What is oolitic marble? Explain the formation of stalactites and stalagmites. What is said of Weyer's Cave? What of the magnesian carbonate of lime?

54. State in full what is said of fluor spar. What is said of apatite? What of the salts of magnesia?

55. What is said of the chlorid of ammonium? What is the composition of common alum? What is said of other alums? What of the *feather alum?* What of the phosphates of alumina?

## CHAPTER VII.

56. What are most of the earthy minerals? What is a silicate? What is said of the compound character of most earthy minerals? What are some of them which are **very** simple? What proportion does silica constitute of the earth's **crust**? What is quartz? What is said of silica in mica and feldspar?

57. What are the properties of silica? What is said of its different forms and conditions? What of its being colored? What is said of flint? What of the three varieties of quartz?

58. What is said of the crystals of quartz? What of rock crystal? What of the amethyst? What of other crystals of quartz?

59. What is chalcedony? What is said of the variety called *carnelian?* What of the *agate?* What of the *onyx?* What of the *cat's eye?* What is the composition of opal? What is said of its varieties? What is said of silica in solution?

60. What are the two states in which silica exists? State in full what is said of the changes from the one state into the other. Explain petrifaction with silica.

61. What is said of the silicates of magnesia? What of the varieties of talc? What of serpentine? What gave this name to it? What is *verd antique?* What is said of chlorite?

62. What is said of pyroxene? What of hornblende? What of its two remarkable varieties, asbestos and amianthus? What is the composition of alumina? What is said of its base? What is emery?

63. What is sapphire? What is said of clay? What of the mineral called spinel? What of the silicates of alumina? What of feldspar? What are its varieties in color and appearance?

64. What is kaolin? What is said of albite? **What of labradorite?** What is **the composition of mica? What are its qualities?** What its uses?

65. What is said of garnet? What of tourmaline? What of the composition of topaz? What of its use in jewelry?

66. What is said of the composition of lapis lazuli? What are **beryls** and **emeralds?** What is said of their size and their value? What of zircon?

## CHAPTER VIII.

76. What is the relation of mineralogy to geology? What is the derivation of the word geology? **In** what senses does the geologist use the term rock? How many elementary substances are there? Name those which enter in any extent into the composition of the rocks. What is said of the amount of oxygen in the earth's crust?

68. What is said of sulphur as a component of rocks? What of

hydrogen? State in full what is said of compounds existing in the rocks. What examples are given of incomplete crystalline structure in rocks?

69. What is said of the structure of sandstone? What of rocks composed of a single mineral? What is a pudding-stone? What a breccia?

70. What is said of stratified and unstratified rocks? Give in full what is said of silicious, argillaceous, and calcareous rocks. How is sulphuric acid used as a test in examining rocks?

71. What is the composition of granite? What is syenite? What porphyritic granite? What is said of the colors of granite? What is said of its dissemination in the earth?

72. What is said of the uses of granite? What of the presence of pyrites in it? What of the selection of granite for building? State in full what is said of gneiss.

73. What is said of the terms slate, shale, and schist? What of mica schist? Of hornblende slate? Of talcose slate? Of clay slate? What is granite rock? What is said of its varieties? What of its uses?

74. State in full what is said about sandstones. What is the meaning of the word trap?

75. What is greenstone? What basalt? What trachyte? What clinkstone? What is said of the term porphyry? What of the term amygdaloid? What form do the trappean rocks tend to assume? What examples are given of this?

76. Describe the ordinary structure of the trappean columns. What is said of the basaltic dike in North Carolina?

77. What of Titan's Piazza? State in full what is said of the trappean rocks in this country. What is lava? What are the two classes of lava?

78. What is scoria? What is pumice? What its uses? What are obsidian and pitchstone? What is said of the resemblance of lavas to the trappean rocks?

## CHAPTER IX.

79. What is said of the substances of which the earth is composed? If the surface of the earth were not diversified, what would be the arrangement of the water and the air? Give in full what is said of the weight of the earth. What estimates are given of the increase of condensation toward the centre?

80. Explain what is represented by Fig. 31.

81. Show how much this figure exaggerates the amount of equatorial bulging. What influence has this bulging upon the earth's rev-

olution? What is meant by the *crust* of the earth? In what ways has knowledge of it to great depths been obtained?

82. What is said of the foundation of this crust? What of its thickness? What is the proportion of land and water on its surface? What is said of the elevations on the earth's crust? What of the depressions? What of the shallows of the ocean? What of bodies of water between certain portions of land?

83. What comparison is made between the equatorial bulging and the mountainous elevations? What between the elevations and depressions? What is said of the ocean near Newfoundland? What of the general arrangement of the land? What of its divisions? What of the position of the Atlantic and Pacific Oceans?

84. What is said of a certain arrangement of the waters noticed by Guyot? What of the arrangement of the mountains?

85. What is said of their relation to the oceans? What of variations in the arrangements of chains of mountains? What as to their prevailing directions? State in full what is said of volcanoes. What is a plateau?

86. What are called lowlands? Give examples of extensive lowlands. What relations have mountains to plateaus and lowlands? What are table-lands? Give examples of extensive plateaus. How do these compare in arrangement with what we often see on a small scale?

87. State in full what is said of rivers. What is said of lakes? How do mountains affect the fertility of regions? How are winds produced? Give what is said of the prevalent winds.

88. State in full what is said of the influence of mountains on the fall of rain, and the consequent difference between America and Europe. What would have been the condition of North America if the Rocky Mountains were on its eastern side and the Appalachians on the western? What is said of the circulation of water?

89. What is said of systematic currents? What comparison is made in regard to the circulation of water? What is said of the diversification of the earth's surface? What of the treasures in the earth's crust?

## CHAPTER X.

90. What is said of the apparent discrepancy between the Scripture account of the creation and the ideas of geologists? In what sense is the word *present* used in the title of this chapter? What are the agents of change in the earth?

91. State in full what is said of the aqueous and igneous agencies. Illustrate the influence of different degrees of rapidity in the flow of water on the removal of materials.

92. State in full what is said of deposits in rivers and on their borders.

93. Explain the mode of improving rivers illustrated by Fig. 35.

94. What is said of deposits in lakes? State the facts in regard to the Lake of Geneva. What is said of the lakes of our country?

95. What is a delta, and why is it so named? Describe the mode of its formation. What is said of bars and islands formed in connection with deltas?

96. Why is there no delta at the mouth of the Amazon? Give in full what is said of the sediment deposited from the Ganges in the Bay of Bengal. What is said of Louisiana, and of the amount of matter brought down by the Mississippi?

97. What is said of the extent of deltas? What of the consolidation of sediment from rivers into rock? Give examples of the encroachment of water upon land.

98. What is said of the mechanical action of water upon rock? What is said of "Pulpit Rock?"

99. Illustrate the erosive work which water does by rubbing one solid against another. What is said of pot-holes? Give the statement in regard to Niagara Falls.

100. What is said of the cañons of Colorado? What of the erosion by the River Simeto? State in full what is said of the *constancy* of the action of water in erosions.

102. How does frost disintegrate rocks? How do the results differ in hard and soft rocks? What is a talus, and how is it formed? What is said of the sorting of fragments in a talus? State in full what is said of the chemical action of water on rocks.

103. Explain weathering. What is said of the Druidical monuments? What of the weathering of hard and of soft rocks? What of weathering in the soil?

104. Give the statement about the boulder. What is said of the Loggan stones? How do the present effects of glaciers and icebergs compare with their effects in past ages? Why is a glacier said to be a river of ice?

105. Of what is a glacier made? What are lateral and medial moraines? Give in full what is said about the termination of a glacier.

107. What is said of the markings and other effects of glaciers?

108. Describe the formation of terminal moraines. How are icebergs produced? What is said of their height? What proportion of an iceberg is above the surface of the water?

109. How extensive are some icebergs? What is stated of their numbers? What is said of their dropping fragments of rock? What of their dragging and stranding?

110. State in full what is said of the leveling operations of water. What is said of the agency of heat?

111. Of what benefit to the earth are volcanoes? What is the distinction between extinct and active volcanoes? How is it supposed that an eruption is produced? Describe the phenomena of an eruption.

112. When was the first eruption of Vesuvius that is on record? What was its condition just before this? State in full what is said of the burial of Herculaneum and Pompeii. What happened to Vesuvius in 1822? What is the size of Etna? What is said of its eruption in 1669?

113. What does Mantell say of one of Etna's eruptions?

114. Describe the crater of Kilauea. Where is Tomboro? Give Lyell's description of the effects of one of its eruptions.

115. Give the facts in regard to Graham's Island. What is said of other volcanic islands?

116. Give the statement about an island near Iceland.

117. How may changes in the temperature of the earth's crust cause earthquakes? Give Prof. Dana's illustration. What is said of the two vibrations produced by an earthquake? State the effects of earthquakes.

118. What is said of solfataras? What of hot springs? Describe the geysers of Iceland.

119. Explain their operation.

120. State in full what is said of sand being moved by wind.

121. What is said of gradual alterations of levels? Give the statement about the Temple of Jupiter Serapis.

122. What does Hugh Miller say about the old coast-lines of Scotland? State the evidences of a gradual change of level in Sweden and Norway.

123. What is said of organic agencies? What of changes made by man?

## CHAPTER XI.

123. What is said of the succession of changes in the construction of the earth's crust?

124. What is said of the development of life upon the earth? Why will what you have learned of present changes help you to understand changes in the past? How does the geologist arrive at his conclusions? By what comparisons as to composition does he determine the origin of rocks?

125. State the case of the shells found in rock. State the observation made about different rocks that contain shells. State the reasoning about the pudding-stone.

126. What observations are made by the geologist in regard to tracks, ripples, etc.? State the case of the slab found in Pottsville. What are the two grand classes of rocks? What is the construction of stratified rocks? Why are they called aqueous rocks? Why fossiliferous rocks?

127. What are metamorphic rocks? What is the derivation of the name? What proof is mentioned that heat is the chief agent of metamorphism? What is said of the appearance and the formation of unstratified rocks?

128. What is the usual difference in position of the stratified and the unstratified rocks? What is said of trappean rocks? What is a stratum? What a stratification? What are laminæ? What is said of them as seen in shales and micaceous sandstones?

129. What is said of the deposition of laminæ? How is the term *formation* used? What are joints and master-joints?

130. What is said of the regularity of these divisions in the rocks? What is stated about the cliffs of Cayuga Lake? What is cleavage in rocks?

131. What is said of the order of succession in the strata? Give the comparison of Phillips. What remarkable fact is stated about the chalk formation? What is said of flexures of strata? What of the circumstances under which they were produced? State the comparison in regard to ice.

132. Illustrate the mode of producing the flexures. What fact bearing on this is related by Lyell?

133. What is said of upheavals of strata accompanied by fracture? What of chasms, caverns, and natural bridges? Illustrate the manner in which strata are brought to view by upheavals.

134. What is true of strata found in a vertical position? Show how strata may be supposed to be horizontal when they are far from it.

135. Explain dip and strike. Show what anticlinal and synclinal lines are.

136. Show in detail how strata may be laid down on a map when they outcrop.

137. Show how the thickness of the strata in the above case may be ascertained.

138. What is said of the extent of knowledge that can be acquired by such observations as have been detailed?

139. Show what are conformable and unconformable strata. What are faults?

140. What is denudation? What is said of the extent of the results effected by it? State the case illustrating denudation occurring with a fault.

141. State the case of the great fault spoken of by Mr. Lesley.

P

142. Point out the different modes in which mountains have been made.

143. Mention the three ways in which valleys have been made. Of what materials are volcanoes constructed? Give the facts in regard to Fusiyama.

145. Indicate the mode of formation of volcanic cones. How are irregularities in their form caused? What is said of the formation of trappean rocks?

146. How is a sunk dike formed? How a raised dike? What comparison is made by Hugh Miller in speaking of the trap dikes about Edinburg? What is said of East and West Rocks in New Haven? What does Hugh Miller say about trap scenery? How were the trap rocks formed?

147. What is said of the commotion attending their formation? What is the statement of Hugh Miller and the criticism upon it? What is said of the pillar-like form often assumed by the trap rocks? What are veins? What is said of veins in granite?

148. State what is said of the granite boulder of which a section is given. How do veins differ from dikes?

149. What is the vein-stone or gangue? What is said of the manner in which dikes and veins are formed? What is a lode? What is drift? In what parts of the continents does it appear? By what means was it transported to its localities? How can it be ascertained where it came from in any case?

150. What traces of the passage of drift are found? What is said of alternate elevations and subsidences of the earth's crust? What prolonged subsidence is spoken of by Lyell? What is said of the subsidences and elevations of the Coal age?

151. By what means are pebbles, sand, and earth produced from rocks? What is soil? What may be said of the additions to it from decay? Is the preparation of soil by disintegration of rock to be regarded only as a destructive process? How may earth-worms and ants be considered as geological workers?

152. Describe the work of the earth-worms. Describe that of the ants.

154. Indicate the manner in which coral reefs are built. To what climates are coral animals confined? Why are there none on the western coast of South America? What is the extent of the reef on the coast of New Holland? Explain the difference between fringing and barrier reefs. How was Florida made?

155. What are atolls and lagoons? Show how an atoll is formed by means of Figs. 81 and 82.

156. What is said of the number of coral islands in the Pacific Ocean? What of the depth to which they reach in the water? What

of the time occupied in their formation? What is said of the formation of limestone rocks from shells?

157. What is said of the foraminifera? What of the agency of minute animals, in comparison with large ones, in forming the earth's crust? What is stated about the nummulites? What is ooze, and what is said of it?

158. Of what are the limestone strata quarried near Paris composed? What did Ehrenberg find in chalk? Mention the observations of Soldani. What is said of silicious shells? What of the white clay about Richmond?

159. What of the tripoli of Germany? What is supposed to be the origin of the flint-stones found in chalk? What is said of the extent of the agency of silicious animalcules and plants in forming rocks? What is the nature of diatoms? What is said of the formation of seas, lakes, and rivers? What is said of plan in the construction of the earth?

160. What is said of life in the different ages of the earth? What of making out the order of succession in the rocks by the life-record?

161. What is said of the resemblances of present living forms to those of the past? What is the nature of the evidence from the life-record as to the relative ages of the strata? What is said of the correspondence of the evidence in different countries? Illustrate this by reference to the coal formation.

162. What are fossils? What is the origin of the term? In what different degrees are fossils preserved? What is said of petrified fossils? What of casts of living substances made by some mineral? Why may coal, strictly speaking, be considered a fossil? State what is said of the preservability of different fossils.

163. Give in full what is said of the abundance of fossils. Explain the principles on which Cuvier and others have investigated fossils.

164. Show how these principles are applied by most people to a limited extent. How is skill acquired in applying them? What is said of their application to vegetables? What is the derivation of the word Palæontology? What is said of the general plan of living structures?

165. What are the Acrogens, and why are they so named? What are the Gymnogens, and what is the derivation of their name? What purpose have these and the Acrogens served?

167. What relations had the Acrogens and Gymnogens to animal life?

169. What are the Endogens? What is their chief purpose?

170. What is meant by a geological, and what by a zoological agency? Explain the meaning of the term Endogens. How does the mode of growth in an Exogen differ from that of an Endogen?

What is the difference in their stems? What is the use of the Exogens? What is said of the presence of the Endogens and Exogens in the life-record? What difficulty is there in the study of fossil botany? What error is mentioned as having risen from the scantiness of the evidence? What is said of the veins in the leaves as furnishing a ground of distinction in the classification of vegetable remains?

173. Why is the study of fossil zoology more satisfactory than that of fossil botany? Give in full what is stated about Protozoans.

174. What are the classes of Radiates? Why are they so called? State what is said of them.

176. Why are Mollusks so called? What does this division include? How far are Mollusks zoological in their relations? Why are they of great value in geological investigations? Why are the Articulata so named? What does this division include? What is said of the relations of the Articulata? What of the variety of their character?

177. What are the characteristics of the Vertebrata?

178. What is said of the spinal marrow of the Vertebrates? What are the four grand divisions of this portion of the animal kingdom? What is said of the intellectual qualities of the Vertebrata? What of their relations, geological and zoological? What of their relations to the higher orders of the vegetable world?

179. What is said of the mode of division of the earth's history? What is Dana's division?

180. Give the division of time into five periods, explaining their names. How definite are the boundaries of the ages? What is said of foreshadowing?

181. What examples of foreshadowing are cited? What is said of the idea on which the ages are named?

## CHAPTER XIII.

181. What is said of the fused state of the earth?

182. What of the time occupied in cooling the earth sufficiently to form a crust? What of the length of time before life appeared on the earth? State in full what is said of the floor of the earth's crust. What were the changes that attended the beginning solidification of the earth's crust?

183. What agency had denudation in this beginning of the formation of land? Give Dana's statement in regard to the land formed in the Azoic age. What is said of the extent of the lands that then appeared above the surface of the waters?

184. What was the shape of North America in the Azoic age? What does Agassiz say of this island? How was Europe, in contrast with North America, in this age?

185. What is said of the thickness of the Azoic strata? What of the upheavals and flexures of them? What of the rocks of this age? What of the temperature of the forming crust?

186. Give Hugh Miller's description of the commotions attending its formation.

187. Why could not life exist in this age? What was the condition of the surface of the land at the end of it?

## CHAPTER XIV.

187. Why is it inferred that life was introduced at the beginning of the Silurian age?

188. Why is it supposed by some that life was introduced before this? What reason have we for saying that life was created? What is said of species? What are the rocks of this age? What is said of the rocks at Niagara Falls?

189. Why is this called the Silurian age? What is the arrangement of the Silurian system in New York?

190. What is the geographical distribution of the Silurian rocks, so far as ascertained? State the origin of the copper of Lake Superior. What is said of the salt springs in this country?

191. State Dana's way of accounting for the great accumulation of salt in the State of New-York. Trace the comparison in regard to the lagoon. How is the salt obtained from the springs in New York?

192. How was the gypsum, which abounds in the Silurian formation in New York, produced? Where gypsum and salt are found together, how are they separated? State in full what is said of "the Silurian beach."

193. What is said of the traces left by water? State in full what is said of life in the Silurian age.

194. What was the condition of the atmosphere in this age? Show how the animal world was adapted to it. Give what Agassiz says of this adaptation. What was the source of the carbonaceous matter in the Silurian shales?

195. What are the Hydrozoa and Bryozoa of this age? What is said of the corals and echinoderms?

196. What of the chain coral? What of the crinoids?

197. Name the three classes of Mollusks, and state what is said of each.

198. What are the Brachiopods, and what is said of them? What is said of the Lingula? Of the Orthoceras? What of the Cephalopods in this age, in contrast with the present?

199. What is said of the Trilobites?

200. What evidence is there that the climate of this age was quite warm over most of the earth? What evidence is there that the sun shone clearly then?

## CHAPTER XV.

201. What were the rocks formed in the Devonian age? Indicate the extent of the Devonian system in this country.

202. What is its extent in Europe? Detail the interesting example of geological investigation mentioned. How prominent were the coral formations in this age? Mention the remarkable display of Devonian corals in Kentucky.

203. What two elegant corals are mentioned as being found in that locality? What different kinds of rock-making were going on at the same time in the Devonian age? What is said of Devonian vegetation?

204. What of Devonian animals? What of Mollusks? What of Trilobites?

205. What of Crustaceans?

206. What is the discovery of Agassiz in regard to the classification of fishes? Of what value is this to geology? Name the four orders of Agassiz, and state what is said of each.

207. What is said of the comparative prevalence of these orders in different ages? Describe the Coccosteus.

208. Describe the Pterichthys. What is said of the appearance of its remains in the rock? Why is the Cephalaspis so named?

209. Describe the Asterolepis.

210. What are *comprehensive* types? What is said of their comparative prevalence in different ages? What is said of the tails of ancient fishes? What of the abundance of fishes in the Devonian age?

211. What of the state in which their remains are often found? What was the condition of the North American continent at the end of the Azoic age? What at the end of the Silurian? What additions were made to it in the Devonian? What results did upheavals in the west and the east produce at the close of this age? What results where the city of Cincinnati now stands?

212. What does Agassiz say of the abundance and richness of Silurian remains found in that locality? What is said of the scenery of the Devonian age?

## CHAPTER XVI.

213. How does the name of the age of coal differ from the names of the other ages? What peculiar propriety is there in the name?

QUESTIONS. 343

214. What is the estimated amount of coal in all the coal-fields of North America? What of all in Great Britain? In Belgium? In France? What facts are stated about the introduction of anthracite into full use in this country? What estimate is made of the length of time the coal of this country will last? What statements are made about the extent of coal-fields?

215. Describe the arrangement of the carboniferous strata. What are called the coal-measures? In what way do the rocky strata in these measures differ from other rocks? Of what practical value is the knowledge of this difference? What does Hitchcock say on this point? What is the *under-clay*, and what is said of it? What is the proportion of coal to rock in the coal-measures?

216. What is the sub-carboniferous period? What is the thickness of the floor of the coal-measures laid in this period? Of what was it made? What is the relative position of the sub-carboniferous and carboniferous strata? What similarity of arrangement was there in the Silurian and Devonian ages and the sub-carboniferous period?

217. In what two respects did this arrangement differ in the sub-carboniferous period from that of the Silurian and Devonian ages? What is the difference between the burning of wood and the making of charcoal? Explain the chemistry of the formation of coal. Why is this process called eremacausis? What is said of the decay of vegetable matter?

218. What is said of the agency of pressure in making coal? What is the difference between bituminous and non-bituminous coal? What is the flame of bituminous coal? What that of imperfectly burning anthracite? How is it supposed that anthracite was produced from bituminous coal? How does it differ from coke? What is said of the vegetable growth from which coal was formed?

219. What is said of the accumulation of material for a coal-bed? Describe what took place after all the material was deposited. What is stated about the number of coal-beds? What does Hugh Miller say of the alternate subsidences and elevations of this age?

220. What is said of the impurities of coal? What of the calculations as to the time required for the formation of the coal-measures? What of the plants whose remains are found in them?

221. What is said of the remains of leaves found in the rocks? State the experiments of Professor Göppert.

222. What is said of the Calamites? What of Sigillaria and Stigmaria?

223. What of the Lepidodendron? What of the Ferns?
224. What of the Conifers?
225. What is the general view given of carboniferous vegetation?
226. What does Page say of the beauty of this vegetation? What

is said of leaf-scars? What was the peculiar condition of the atmosphere up to the Carboniferous age?

227. What change was made in it during that age? How was it done? State the comparison made in regard to this transfer from the gaseous to the solid state. What is said of the climate of this age? What was the character of Carboniferous scenery?

228. What is said of the extent of the successive platforms of vegetation? What change occurred in the scene when the accumulated material was submerged for conversion into coal? What is said of the animals of this age?

229. What is said of the coral animals? What of the crinoids and sea-urchins?

231. What of the Crustaceans? What of the Reptiles? State the facts about the Pottsville reptile. What is said of the fishes?

232. State in full what is said of the Lepidosteus and of the garpike of the present age. What is said of the Mollusks? What of the Productus spinulosus?

233. What of the Spirifer? From what does the Permian period get its name? What is said of the Permian strata in this country? What of them in Europe?

234. What are the Permian rocks in this country? What is said of the Permian limestones of Europe? What was the condition of North America at the end of the age of Coal? What is said of the disturbances of the coal-measures?

235. What is said of the connection of metamorphosis with these disturbances? What of their being systematic? What of their rate of movement? What of certain valuable results of metamorphosis at this time? What of the debituminization of coal? What of denudation? What of the differences of the forms of life in different ages?

236. When did vertebrates first appear? When reptiles? What class of animals did not appear till after Palæozoic time? What is said of the disappearance of animals? What of the destruction of life at the close of the Carboniferous age?

## CHAPTER XVII.

236. What are the divisions of the age of Reptiles? Whence comes the term Triassic? Whence Jurassic?

237. What is said of the term Oolitic? What of the term Cretaceous? What is stated about the Cretaceous system in this country? State in full what is said of the Triassic rocks. What is said of the localities in which they are the surface rocks?

238. Why is this system often called in Europe the Saliferous system? How is the production of salt there in contrast with its pro-

duction in this country? What is said of Triassic plants? What of Triassic coal? How did the atmosphere differ from what it was up to the Carboniferous age? What animals were consequently now introduced?

239. What kinds of mammals were introduced? What kinds of fishes? What is said of the Mollusks? What is the Lily Encrinite? What is said of the Reptiles? Describe the Labyrinthodon.

240. What is said of the name Cheirotherium? What of the teeth of the Labyrinthodon?

241. What of the tracks of Triassic animals? What of the slab represented in Fig. 141? What of the Brontozoum giganteum?

242. What of the track represented in Fig. 143? What of the stone fossil book?

243. What are some of the notable localities of trap rocks in this country? As they were thrust up, what effect resulted in the adjacent rocks? What is said of their systematic arrangement?

245. What is said of the vegetation of the Triassic period? State the facts about the Portland dirt-bed.

246. Give in full what is said of the animals of this age.

247. What were the most prominent of the mollusks of this period? What is the class to which they belong, and what are the characteristics of the class?

248. What is said of the Ammonites? What of the Belemnites? What of the Ichthyosaurus?

249. What of the Plesiosaurus?

250. What of the Pterodactyl? What has been the common opinion about the nature of this animal? What is Agassiz's opinion, and what are his reasons for it?

252. What is probably the truth about the flying of the Pterodactyl? What is said of the Dinosaurs?

253. What are the rocks of the Cretaceous system in Europe and Asia? What in this country? What is said of the *Green-sand?* What of chalk? What of the localities of the Cretaceous system?

254. How was chalk formed? Of what is it constituted as shown by the microscope? What is said of chalk-marks? What is the actual difference between chalk and other forms of carbonate of lime? What is said of deposits of chalk now going on?

255. What is said of the presence of flint in chalk? What of the Xanthidia?

256. What of the shell prisms in flint? What of the flinty spicula from sponges?

258. What of the agency of sponges in the Cretaceous period? What of the animals of that period? What of the change in the forms of the Ammonites?

P 2

259. What is said of the formation of mountains at the close of the Reptilian age, and in the age after it? Give in full what is said of the evidence in regard to this.

260. What is stated about the heights at which animal remains are found in the mountains raised at that time? What is said of the destruction of life at the end of this age? What is Hugh Miller's division of the life of the earth? What is said of new creations?

## CHAPTER XVIII.

261. **In the age of Mammals**, how was the condition of the earth more like its present condition than it was in the previous ages? What is meant by Cainozoic time? What are its divisions?

262. What is said of the terms primary, secondary, and tertiary? Give in full what is said of Lyell's division. Of what kinds, mostly, were tertiary deposits?

263. What is said of **Tertiary** mediterranean seas? State in full what is said of the Paris basin. What is said of the areas of Tertiary deposits?

264. **What are the Tertiary rocks?** What is said of some of the Tertiary strata of Europe? What is said of Nummulites?

265. **State in full** what is said of the **Nummulitic formation.**

266. What is said of Tertiary plants? Give in full what is said of diatoms. What is said of Tertiary animals?

267 What comparison is made between Tertiary animals and those of the present age? What is said of the fishes?

268. What of the reptiles? **What of the mollusks?** Give the statement about the arrangement of a cliff in Virginia.

269. State some of the inferences drawn from the condition of shells in strata. State in full what is said of indusial limestones.

270. What is said of the Mammals of this age?

271. What is said of the whale called Zeuglodon cetoides?

272. To what class did the most prominent of Tertiary mammals belong? Describe those which are represented in Fig. 164. What comparison is made between the pachydermata of the present time and those whose remains were found in the Paris basin?

273. What is said of Cuvier's investigations? To what conclusion did he come? How were his views received? State in full what is said of the Dinotherium.

275. Indicate the extent of Tertiary mountain-making. Give in full the comparison between the **Tertiary** mountains and those made in earlier ages.

276. What is said of systems of rivers? What evidences are there of the working of volcanic agencies in the Tertiary period? What is the explanation of the distinctness of some of the cones in France?

277. What is the name of the highest of these cones, and what is said of it? How were basins, so called, formed in the Tertiary period? What is the arrangement of the London basin?

278. How do artesian wells operate in such basins? What is said of Tertiary continent-making? What proportion of the land was brought above the surface of the water in this period? What additions were made to North America? What is said of the condition of this continent at the end of this period?

279. What is the Post-tertiary period? What was to be done in it to the continents? Into what three epochs is it divided? What was done in the first? What in the second?

280. What in the third? Indicate the extent of the reign of ice in the Glacial epoch. What is supposed to have been the cause of the cold? What is said of the glaciers of this period?

281. What is drift? What modified drift? What are the two theories about drift? What is the truth in regard to them? What changes were produced by the drift? What change in the Niagara River? What is said of the heights at which drift is found?

282. What is said of the shape and size of boulders? What of the distances to which they have been carried? What examples are given of the transportation of them across deep valleys? What is stated about the distribution of boulders?

283. What is said of glacial markings? What of the *roches montonneés*? What of the work done by icebergs, and that done by glaciers? What of the localities of the glacial markings?

284. What general change in the land occurred in the Champlain epoch? What resulted from this change? What is the character of the formations of this epoch? How are boulders situated in relation to these formations? What is said of the elevated heights to which these formations sometimes reached? What is said of the wide dispersion of the smaller materials of the drift?

285. What are the river-plains and sea-beaches of the Champlain epoch now? What is said of the subsidence of land in this epoch? What of the movement in the Terrace epoch? What effect was produced by this on the Champlain strata? To what is the term terraces applied? What is said of the arrangement of terraces?

286. Describe what is represented in Figs. 168 and 169. State in the general how terraces were formed.

287. Explain by Figs. 170 and 171 the manner in which they were commonly made.

288. What is said of sea-beaches? Describe in full the changes that occurred in the Niagara River in the Post-tertiary period.

289. What calculations have been made about the time occupied

in the recession of Niagara Falls? How are rivers formed? What two things are needed to make large rivers?

290. When were the grand river systems of the earth formed? What is said of the alterations in the terminal moraines of the ancient glaciers? What of the results obtained in their investigation?

291. What cities are built on moraines? What is the composition of the Post-tertiary moraines? What is said of the present concealment of their character? What is soil? By what agencies has it been made? What is said of their operation in the Post-tertiary period?

292. What does Hugh Miller say of the preparation of soil in Scotland in that period?

293. What three points are worthy of remark in regard to Post-tertiary animals? What is said of the animals of England of that period?

294. How did the Siberian mammoth differ from the elephants of the present time?

295. Give in full what is said of the covering of the mammoth in relation to the question of climate. What is said about vegetation in reference to this? What is said of the abundance of mammoths in certain localities, and of the supply of ivory from their tusks? What is stated about the mammoths of this country?

296. Give the history of the St. Petersburg skeleton. State the speculations about it, and the views of Cuvier.

297. What is said of the mastodons? How do their teeth differ from those of the mammoth? What is said of the Mastodon giganteus?

298. What is the Indian tradition about mastodons? What is stated about the Newburg mastodon? What is said of the Mylodon?

299. What is said of the structure of the Mylodon's skull?

300. What is said of the Megatherium?

301. What of the Megalonyx?

302. What of the Glyptodon? What of the whales of that period?

303. Where have caves been found containing bones of Post-tertiary animals? Describe the Kirkdale cave. What was the supposition of some quarrymen about it?

304. State in full the results of Dr. Buckland's examination of this cave.

## CHAPTER XIX.

305. What is said of the beginning of the age of Man? What about its conclusion?

306. Point out the contrast between the earth now and in the Azoic age. What is said of the contrast between the present age and the

ages following the Azoic? How was the Post-tertiary period the most like the present?

307. What facts show that the earth was essentially completed when man was introduced? What is said of his control over the earth? Give the summary of the changes now going on in the earth.

308. What is said about the preservation of the equilibrium of land and ocean at the present time? What of the comparative activity of agents of change now and in former ages?

309. What is the distinction between the historic and the human period? What is said of history? What of the study of the remains of a people? What is said of the *Stone* age of man? What of the specimens of stone implements represented in Fig. 179? What of the *Bronze* age?

310. What of the *Iron* age? What of literature in relation to these ages? What change was made in regard to the animals of the earth when the age of Man began?

311. Mention some of the animals that have become extinct during the age of Man, and state what is said of them. What is said of the evidence that man is one species?

313. Of what nature is the grand difference between man and other animals? In what respects is man like other animals, and why? What power has the human mind which the brute mind has not? Mention the various results of the possession of this power. What supposition has been entertained by some in regard to future progress? How is this seen to be unfounded?

## CHAPTER XX.

314. Give the comparison between the study of Astronomy and that of Geology.

315. What is said of the comparative reliability of history and the record of the rocks? What erroneous idea was entertained by some in the infancy of geology? What is said of the authenticity of the two records of creation? What of apparent inconsistencies between them? What of the proper mode of interpreting the Mosaic record?

316. What is said of the coincidences between the two records? What is said to show that the account of Moses was divinely inspired? What is said of the propriety of noticing the record of the Bible in a work on geology? What were the *days* of the Mosaic record? What objection to the view taken is mentioned, and what is said about it?

317. What are the various senses in which the word day is used in the Mosaic account? What significant fact is stated about the use of the word? Relate the English legend about ammonites.

318. Relate the imposition practiced in regard to this legend. Of

what use is geology to the poet and the painter? What is said of a plan in the formation of the earth?

319. Give the comparison in regard to the potter. Illustrate the fact that the continents were constructed each on its own plan. What facts show that the Creator had a plan throughout in regard to the animals of the earth? What is said of *time* in geological processes?

320. What is said of minute agencies in the formation of the earth's crust? What of disintegration of the rocks? What of the extent to which reconstruction has been carried on in the formation of the earth's crust?

321. What have been the two purposes of disintegration? Which is the principal one? How has diversification of the earth's surface been effected? Show the interplay of mechanical, chemical, and vital forces in the case of carbonate of lime. What was the state of the atmosphere before the Carboniferous age? What change was effected in it then, and with what result?

322. Show how the interplay above referred to is illustrated in the making of coal in the Carboniferous age. What is said of the circulation of matter? What has been the notion of some in regard to the creation of the earth? What is the truth on this point?

323. What is said of the creation of life? What of the notion that there has been no creation at all? State the development theory. How does geology show this to be false?

324. How does Page state the changes implied in this theory? What is the grand creation of the Deity on this earth? What two facts show that the soul of man was a distinct creation?

325. If mind were entirely dependent upon matter, what ought we to find in the brain of man? What is said of the doctrine that places man only a little above the monkeys?

# GLOSSARY.

[The numbers refer to the paragraphs where the terms are explained.]

| Term | ¶ | Term | ¶ |
|---|---|---|---|
| Acephal | 286 | Faults | 223 |
| Acrogens | 250 | Ferruginous | 113 |
| Amorphous | 21 | Flux | 69 |
| Amphigens | 249 | Fossiliferous | 211 |
| Amygdaloid | 152 | Fossils | 245 |
| Anhydrous | 60 | Gangue | 69 |
| Anticlinal | 219 | Ganoid | 299 |
| Aqueous (rocks) | 211 | Gasteropod | 286 |
| Arenaceous | 145 | Gymnogens | 251 |
| Argillaceous | 145 | Haloid | 95 |
| Atoll | 238 | Hydrocarbon | 46 |
| Azoic | 261 | Indusial | 379 |
| Bilobed | 302 | Joints (in rocks) | 212 |
| Bort | 35 | Jurassic | 339 |
| Brachiopod | 286 | Lacustrine | 369 |
| Cainozoic | 261 | Lagoon | 238 |
| Calcareous | 145 | Lamination | 212 |
| Cenozoic | 261 | Lode | 231 |
| Cephalopod | 286 | Massive (minerals) | 105 |
| Cleavage | 24, 212 | Master-joints | 212 |
| Coal-measures | 310 | Mesozoic | 261 |
| Comprehensive (type) | 301 | Metamorphic | 211 |
| Conformable (strata) | 222 | Miocene | 368 |
| Cretaceous | 339 | Nebulous state | 263 |
| Ctenoid | 299 | Nummulite | 239 |
| Cycloid | 299 | Octahedron | 13 |
| Deliquesce | 97 | Oolite | 102 |
| Denudation | 224 | Organized (substances) | 3 |
| Devonian | 290 | Outcropping | 220 |
| Dimorphous | 21 | Palæontology | 247 |
| Dip (of strata) | 218 | Placoid | 299 |
| Dodecahedron | 25 | Pliocene | 368 |
| Dune | 205 | Primary (forms) | 25 |
| Endogens | 252 | Protozoans | 256 |
| Eocene | 368 | Rhomb | 25 |
| Eremacausis | 313 | Saliferous | 278 |
| Exogens | 253 | Secondary (forms) | 26 |

| | | | |
|---|---|---|---|
| Silurian | 275 | Trappean | 152 |
| Stratification | 212 | Trend | 162 |
| Strike of strata | 218 | Triassic | 339 |
| Sub-carboniferous | 311 | Tufa | 195 |
| Synclinal | 219 | Unilobed | 302 |
| Talus | 185 | Water of crystallization | 20 |
| Tertiary | 368 | Weathering | 187 |

# INDEX.

[The numbers refer to the paragraphs.]

Acephals .......................... 286
Acrogens ......................... 250
Adaptation ....................... 282
Æpiornis maximus ............ 421
Agate.. ........................... 114
Ages, boundaries of ............ 262
**Ages** of Man, Stone, Bronze, and Iron........................ 420
Aiguille de Dree ................ 146
Alabaster......................... 60
**Albite** ............................... 129
Algæ ............................... 249
Alum............................... 109
Alumina........................... 126
Alumina, phosphates of........ 110
Alumina, silicates of ........... 128
Amber ............................. 45
Amethyst ......................... 113
Amianthus........................ 125
**Ammonites**........... 344, 352, 363
Ammonium, chlorid of........ 108
Amorphous minerals ........... 21
Amygdaloid ...................... 152
Anoplotherium .................. 382
**Anticlinal** lines .................. 219
**Antimony** ........................ 84
Antimony, sulphuret of........ 54
Anthracite ........................ 39
Anthracite, how produced.... 315
Anthracite, introduction of... 308
Ants, as geological workers.. 235
Aqueous rocks ................... 211
Arenaceous rocks ............... 145
Argillaceous rocks .............. 145
Arsenic, sulphurets of.......... 55
Articulata................... 259, 332
**Asbestos**.......................... 125

**Asphaltum**....................... 46
Asterolepis....................... 301
Astronomy, Geology compared with ..................... 425
Atmosphere, relation of to coal............................... 437
Atolls .............................. 238
Auvergne, volcanoes of ....... 386
**Azoic** time........................ 261
**Azoic** age, heat and light in 271
**Azoic** age, land of........266, 267
Azoic age, rocks of ............ 270

Barrier reefs..................... 237
Baryta, sulphate of............. 62
Basalt.............................. 152
Basilosaurus..................... 381
Basins............................. 387
Belemnites....................... 352
Beryls ............................. 135
Bismuth........................... 87
Bitumen.......................... 46
Black Jack....................... 57
Boracite .......................... 107
Borax .............................. 99
Boulders.......................... 393
Brachiopods ..................... 286
Breccia ............................ 143
Brontozoum giganteum....... 346
Bryozoa ........................... 284
Buhrstone........................ 150

Cainozoic time ................... 261
Cainozoic time, divisions of.. 368
Calamites ........................ 32
Calcaire grossier ............... 369
Calcareous rocks................ 145

| | |
|---|---|
| Calcareous spar | 101 |
| Carbon, diffusion of | 29 |
| Carbonic acid | 47 |
| Carbureted hydrogen | 48 |
| Carcharodon | 376 |
| Carnelian | 114 |
| Caverns | 216 |
| Cenozoic time | 261 |
| Cephalaspis | 300 |
| Cephalopods | 286 |
| Chain coral | 285 |
| Chalcedony | 114 |
| Chalk | 100 |
| Chalk, flint in | 361 |
| Chalk, source of | 360 |
| Chalk, what composed of | 239, 383 |
| Champlain epoch | 395 |
| Changes in the earth, agents of | 172 |
| Cheirotherium | 345 |
| Chlorite | 123 |
| Chrome yellow | 75 |
| Chrysoberyl | 135 |
| Cleavage | 24, 212 |
| Clink-stone | 152 |
| Coal | 38 |
| Coal, origin of | 41 |
| Coal, diffusion of | 44 |
| Coal, localities of | 307 |
| Coal, how made | 313 |
| Coal, how deposited | 316 |
| Coal, plants in | 319 |
| Coal, rate of formation of | 318 |
| Coal, why called a mineral | 42 |
| Coal age, vegetation of | 325, 326 |
| Coal age, climate of | 327 |
| Coal age, scenery of | 328 |
| Coal age, animals of | 329, 334 |
| Coal age, North America at close of | 336 |
| Coal-fields | 309 |
| Coal-measures | 310 |
| Coal-measures, disturbances in | 337 |
| Coast-lines of Scotland | 207 |
| Cobalt | 85 |
| Coccosteus | 300 |
| Colorado, cañons of | 184 |
| Columnar trap | 153, 154 |
| Confervæ | 249 |
| Conformable strata | 222 |
| Continent-making in the Tertiary period | 388 |
| Copper, native | 77 |
| Copper, oxyds of | 78 |
| Copper, carbonates of | 79 |
| Copper, silicate of | 80 |
| Copper of Silurian rocks | 277 |
| Coral reefs | 237 |
| Corals | 236 |
| Corals in age of Coal | 330 |
| Corals in Cretaceous period | 363 |
| Corals in Devonian age | 293 |
| Corals in Jurassic period | 351 |
| Corals in Silurian age | 285 |
| Creation, the two records of | 427 |
| Creative power | 439 |
| Cretaceous period, animals of | 363 |
| Cretaceous system | 358, 359 |
| Crinoids in Coal age | 331 |
| Crinoids in Cretaceous period | 363 |
| Crinoids in Jurassic period | 351 |
| Crinoids in Silurian age | 285 |
| Crust of the earth, floor of | 264 |
| Crust of the earth, treasures in | 170 |
| Crystallization | 13 |
| Crystallization, contrasted with vital growth | 15 |
| Crystallization, modes of | 19 |
| Crystallization, water of | 20 |
| Crystals, arrangements of | 16, 22 |
| Crystals, primary forms of | 25 |
| Crystals, secondary forms of | 26 |
| Crystals, constancy in forms of | 27 |
| Crystals, symmetry in | 28 |
| Crystals, sizes of | 14 |
| Ctenoids | 299 |
| Cuvier, investigations of | 247, 383 |
| Cycloids | 299 |
| Decay | 6 |
| Deliquescence | 97 |

# INDEX.

Deltas..................177, 179
Denudation.................224, 225
Deposits, consolidation of.... 180
Development theory........... 440
Devonian age, corals in...... 293
Devonian age, crustaceans in 298
Devonian age, mollusks in... 297
Devonian age, vegetation in 295
Devonian age, abundance of fishes in...................... 303
Devonian age, scenery of.... 305
Devonian age, rocks of........ 290
Devonian age, United States at end of.................... 304
Devonian system..........291, 292
Diamonds.....................30–36
Diatoms....................240, 374
Dikes.......................... 229
Dimorphous minerals......... 21
Dinornis elephantoides....... 421
Dinosaurus.................... 357
Dinotherium................... 384
Dip............................ 218
Dodo........................... 421
Dolomite...................... 104
Drift......................232, 392
Druidical monuments......... 187
Dunes.......................... 205
Dutch white................... 63

Earth as a whole.............. 156
Earth, form of................ 157
Earth, crust of............... 158
Earth, ages of................ 171
Earth, completion of.......... 416
Earth now and in former ages 415
Earth, stages in construction of............................ 209
Earth, plan in construction of 242
Earth, solidification of........ 263
Earth's surface, elevations and depressions in.....160, 233
Earth's surface, how diversified............................ 169
Earth development, plan in.. 431
Earth's crust, floor of.......... 264
Earth's history, divisions of.. 261

Earthquakes..............200, 201
Earth-worms as geological workers....................... 235
Ehrenberg on chalk........... 239
Elevations..................... 233
Emeralds...................... 135
Emery......................... 126
Endogens...................... 252
England, post-tertiary animals of........................ 404
Eocene......................... 368
Epsom salts.................... 61
Eremacausis................... 313
Etna........................... 196
Exogens........................ 253

Faults......................223, 225
Feldspar....................... 129
Ferns in coal strata........... 323
Fingal's Cave.................. 153
Fishes......................... 299
Fishes of age of Coal.......... 333
Fishes of Tertiary period..... 376
Fishes, tails of................ 302
Flexures of rocks.............. 214
Flint, fossils in................ 362
Flint in chalk.................. 361
Flood-plain.................... 174
Fluor spar..................... 105
Fluxes......................... 69
Foraminifera................... 239
Fool's gold..................... 50
Formation...................... 212
Fossils.....................245–247
Fossils, animal................ 255
Fossils, vegetable.............. 254
Fossil book in stone........... 347
Fossiliferous rocks............. 126
Franconia Notch................ 395
Fringing reefs.................. 237
Fungi.......................... 249

Galena......................... 52
Gangues........................ 69
Ganoids........................ 299
Garnets........................ 131
Gar-pike....................... 333

# INDEX.

Gasteropods ...... 186
Geological evidence, nature of ...... 210, 244
Geysers ...... 203, 204
Giant's Causeway ...... 153
Glacial period ...... 368, 391
Glacial markings ...... 394
Glaciers ...... 188–190
Glance, copper ...... 51
Glauber's salt ...... 61
Gneiss ...... 148
Graham's Island ...... 199
Granite ...... 146, 147
Graphite ...... 37
Green-sand ...... 358
Green-stone ...... 152
Gold ...... 91–93
Göppert, experiments of ...... 319
Grotto del Cane ...... 47
Gulf Stream ...... 168
Gymnogens ...... 251

Haloid salts ...... 95
Hamburg white ...... 63
Heat, agency of ...... 193
Heat, relation of to forms of matter ...... 2
Hematite ...... 74
Herculaneum ...... 195
Historic period ...... 419
Hornblende ...... 125
Human period ...... 419
Hydrozoa ...... 284
Hylæosaur ...... 357

Icebergs ...... 191
Iceland spar ...... 101
Icononzo, natural bridge at ...... 216
Ichthyosaurus ...... 353
Iguanodon ...... 357
Indusial limestones ...... 379
Iridium ...... 94
Iron, ores of ...... 70
Iron, meteoric ...... 71
Iron, carbonate of ...... 76
Iron mountains ...... 73
Islands, volcanic ...... 199

Joints ...... 212
Jupiter Serapis, Temple of ...... 206
Jurassic period, plants of ...... 349
Jurassic period, animals of ...... 351

Kaolin ...... 129
Kilauea, Crater of ...... 197
Kirkdale Cave ...... 413

Labradorite ...... 119
Labyrinthodon ...... 345
Lagoons ...... 238
Lakes ...... 166
Lakes, deposits in ...... 176
Lamination ...... 212
Land, arrangement of ...... 161
Land, proportion of to water. 159
Lapis lazuli ...... 134
Laurentian Hills ...... 385, 388
Lead ...... 81
Lead, oxyds of ...... 81
Lead, sulphuret of ...... 52
Lead, sulphate of ...... 59
Lepidodendron ...... 322
Lepidosteus ...... 333
Levels, alterations of ...... 206
Life introduced by the Creator ...... 15
Life in different ages ...... 243
Life, dawn of, in the earth ...... 273
Life, plan in structures of ...... 248
Life-record in the rocks ...... 243
Life-record, Palæozoic ...... 338
Lily Encrinite ...... 344
Lime, carbonate of ...... 100
Lime, silicates of ...... 119
Limestone ...... 100
Lingula ...... 286
Lisbon, earthquake at ...... 201
Lithodomes ...... 207
Loggan stones ...... 187
London basin ...... 387
Lowlands ...... 164

Magnesia, salts of ...... 107
Magnesia, silicates of ...... 120
Magnesia, sulphate of ...... 61

# INDEX.

Magnesian carbonate of lime 104
Magnetic iron ore .............. 71
Mammals ...................... 260
Mammals of Tertiary period 380
Mammals of Triassic period 344
Mammoth Cave of Kentucky 16
Mammoths ..................... 405
Man, creation of .............. 441
Man, changes made by ........ 208
Man one species ............... 422
Man's place in nature ......... 423
Manganese ..................... 88
Marble ........................ 102
Master-joints ................. 212
Mastodons ..................... 408
Matter, circulation of ........ 438
Matter, forms of .............. 2
Megalonyx ..................... 410
Megalosaur .................... 357
Megatherium ................... 410
Mercury ....................... 89
Mercury, sulphate of .......... 56
Metamorphic rocks ............. 211
Mesozoic time ................. 261
Mica .......................... 130
Microscopic animals in rock 339
Minerals, forms of ............ 1
Minerals in living substances 17
Minerals, simple and compound ... 9
Mines, salt ................... 98
Minium ........................ 81
Miocene ....................... 368
Mollusks ...................... 258
Mollusks in age of Coal ....... 334
Mollusks in Cretaceous period ... 363
Mollusks in Devonian age ...... 297
Mollusks in Silurian age ...... 286
Mollusks in Tertiary period .. 378
Moraines ............. 188, 190, 402
Moraines, cities built on ..... 402
Mosaic record, authenticity of ... 427
Mosaic record, days in ........ 428
Mountain-making in Tertiary period ... 385

Mountains, arrangement of .. 162
Mountains, how made .......... 226
Mountains, relation of, to fertility ... 167
Mylodon ....................... 409

Naphtha ....................... 46
Nautilus ...................... 286
**Niagara Falls, recession of** ... 183
Niagara River, changes in 392, 399
Nickel ........................ 86
Nummulites .................... 239
Nummulitic formation .......... 372

Oaze .......................... 239
Obsidian ...................... 155
Olivine ....................... 152
Onyx .......................... 114
Oolite ........................ 102
Opal .......................... 115
Ores ........................ 66–68
Organic agencies .............. 207
Organized substances .......... 3
Orpiment ...................... 55
Orthoceras .................... 286
Osmium ........................ 94
Oxysalts ...................... **95**

Pachydermata of Tertiary period ... 382
Palæotherium .................. 382
Palæozoic time ................ 261
Palisades ..................... 154
Peat .......................... 43
**Pebbles** .................... 234
Permian period ................ 335
Petrifactions ................. 118
Petroleum ..................... 46
Pitchstone .................... 155
Placoids ...................... 299
Plateaus ...................... 164
Platinum ...................... 48
Plesiosaurus .................. 354
**Pliocene** ................... 14
Pompeii ....................... 195
Porphyry ...................... 152
Portland dirt-bed ............. 350

Post-tertiary period....... 368, 389
Post-tertiary period, divisions of .............................. 390
Post-tertiary period, animals of ................................. 404
Post-tertiary period, length of 400
Post-tertiary period, rivers of 401
Potosi, situation of............. 164
Primary forms of crystals.... 25
Protogine........................... 146
Protozoans ....................... 256
Pterichthys ...................... 300
Pterodactyl ................. 355, 356
Pudding-stones ................ 143
Pulpit Rock...................... 182
Pumice ............................ 155
Pyramids, of what rock built 239
Pyrites, iron..................... 50
Pyrites, copper.................. 51
Pyroxene.......................... 124

Quartz............................. 113
Quartz rock...................... 150
Quito, situation of .............. 164

Radiates ......................... 257
Realgar ........................... 55
Reconstruction.................. 435
Red lead........................... 81
Reptiles of age of Coal........ 333
Reptiles of Tertiary period .. 377
Reptiles of Triassic period ... 345
Reptilian age, destruction of life at close of................. 365
Reptilian age, uplifts at close of .................................. 364
Rhodium ......................... 94
Richmond, white clay of...... 240
Ripple marks ..................... 210
Ripple marks on Silurian beach............................. 280
Rivers, amount of deposits from............................. 178
Rivers, deposits in............. 174
Rivers, improvement of ....... 175
Rivers of post-tertiary period 401
Rivers, systems of ............. 165

Roches montonneés............ 394
Rock, definition of ............ 138
Rocks, disintegration of....... 434
Rocks, elementary substances in ................................. 139
Rocks, kinds of ...... 140–145, 211
Rocks, order of succession of 213
Rocks, regularity of form in.. 23
Rubies....................... 126, 127

Salt mines........................ 98
Salt of Triassic system........ 342
Saltpetre.......................... 96
Salt springs...................... 98
Sand............................... 234
Sand moved by wind .......... 205
Sandstones....................... 151
Sapphire.......................... 126
Satin spar .................. 60, 101
Schists ............................ 149
Scoria............................. 155
Sea-beaches ..................... 398
Secondary forms of crystals.. 26
Serpentine ....................... 122
Shales............................. 149
Shells in rocks....... 239, 246, 258
Shells, silicious.................. 240
Sigillaria ......................... 321
Silica.............................. 112
Silica in solution ............... 116
Silica in two states ............. 117
Silica, petrifactions with...... 118
Silicates .......................... 111
Silicious rocks................... 145
Silurian age, climate of....... 288
Silurian age, life in............. 281
Silurian age, subsidences and elevations in ................... 289
Silurian beach .................. 280
Silurian copper .................. 277
Silurian salt ..................... 278
Silurian shales, carbonaceous matter in......................... 283
Silurian system, distribution of .................................. 276
Silurian system, rocks of..... 274
Silver............................. 90

# INDEX. 359

Silver, sulphuret of ............ 52
Slates .............................. 149
Smalt............................... 85
Soapstone ........................ 121
Soda, borate of................. 99
Soda, nitrate of................. 97
Soda, sulphate of .............. 61
Soil made in post-tertiary period................................ 403
Soil, what it is .................. 234
Solfataras ........................ 202
Sphinx, what made of........ 239
Spicula in flint nodules....... 362
Spinel.............................. 127
Stalagmites and stalactites... 103
Stigmaria......................... 321
Strata .............................. 212
Strata at different angles..... 217
Strata, conformable and unconformable ................. 222
Strata, flexures of .............. 214
Strata, mapping of............. 220
Strata, measuring of .......... 221
Strata, upheavals of........... 216
Stratification .................... 212
Stratified rocks................. 144
St. Petersburg skeleton ...... 406
Strike .............................. 218
Sub-carboniferous period .... 311
Subsidences...................... 233
Sugar, crystallization of...... 18
Sulphur............................ 49
Sulphurets ................50–57
Sulphureted hydrogen ........ 64
Sulphuric and sulphurous acids ............................ 63
Syenite ............................ 146
Synclinal lines .................. 219

Talc ................................ 121
Talus............................... 126
Terrace epoch ................... 396
Terraces, how formed ........ 397
Tertiary period ................. 368
Tertiary period, deposits of.. 369
Tertiary period, mountains made in ...................... 385

Tertiary period, pachydermata of......................... 382
Tertiary period, plants of..... 373
Tertiary period, rocks of...... 371
Time in geology................ 432
Tin .................................. 82
Tomboro .......................... 198
Topaz......................126, 133
Tourmaline....................... 132
Tracks as evidences in geology.. .... ........................ 210
Tracks in Triassic rocks...... 346
Trachyte .......................... 152
Trap rocks............152–154, 348
Trap rocks, how formed..229, 230
Trap scenery ..................... 229
Triassic period ................. 339
Triassic rocks ................... 330
Triassic plants .................. 343
Triassic salt ..................... 342
Triassic animals................. 344
Trilobites......................... 287
Trilobites of age of Coal...... 332
Tufa................................ 195
Turquois .......................... 119

Unconformable strata......... 222
Under-clay....................... 310
Unstratified rocks........144, 211

Valleys, kinds of................ 227
Venice white..................... 63
Ventriculites ..................... 361
Veins............................... 231
Verd-antique..................... 122
Vertebrates...................... 260
Vesuvius.......................... 195
Viesch, glacier of .............. 188
Vitriols ............................ 58
Volcanoes...................163, 194
Volcanoes, formation of...... 228
Volcanoes of the Tertiary period............................. 386

Water as a leveler ............. 192
Water changing locality of materials....................... 173

Water, chemical action of.... 186
Water, circulation of.......... 168
Water encroaching on land.. 181
Water, erosive power of..182, 184
Water, proportion of land to 159
Weathering...................... 187
Whales of post-tertiary period 412
White lead........................ 81

Xanthidia ........................ 362
Zaffre............................. 85
Zeuglodon cetoides ........... 381
Zinc .............................. 83
Zinc, sulphuret of.............. 57
Ziphodon........................ 382
Zircon........................... 136

THE END.

www.ingramcontent.com/pod-product-compliance
Lightning Source LLC
Chambersburg PA
CBHW031426230426
43668CB00007B/449